CIVIL ENGINEERING HERITAGE

Wales and Western England

Edited by W. J. SIVEWRIGHT
OBE, MA, CEng, FICE

THOMAS TELFORD, LONDON

Published for the Institution of Civil Engineers by Thomas Telford Ltd,
Telford House, PO Box 101, 26–34 Old Street, London EC1P 1JH,
England

First published 1986

British Library Cataloguing in Publication data:
Civil engineering heritage: Wales and western England
   1. Civil engineering – England – Midlands
   2. Midlands (England) – Public works – Guide-books
   I. Sivewright, W.J.
   624'.09424   TA58.M/

ISBN: 0 7277 0236 X

Typeset by MHL Typesetting Ltd, Coventry
Printed and bound in Great Britain by Redwood Burn Limited, Trowbridge, Wiltshire

Acknowledgements

The principal contributors to this book are Panel members P. Dunkerley, R. V. Hughes, R. Cragg, O. M. Gibbs and the Editor, W. J. Sivewright.

A number of items were also drafted by the Panel's Vice-Chairman, M. F. Barbey, without whose guidance and criticism this book would not have been possible. He set out the content and general scope of the work and has freely made available his wealth of knowledge of civil engineering history and his experience in editing the first volume of the series.

The Panel's Technical Secretary, T. B. O'Loughlin, has performed sterling service in checking the text, verifying references (in which he has been aided by M. Chrimes of the Institution's Library staff), getting together the illustrations and seeing the book through the various procedures leading to publication.

The Chairman of the Panel, Professor A. W. Skempton, has given valuable advice in the later stages of preparing the book for publication.

Thanks are due to various local authorities, public utilities and others, too numerous to mention in detail, for their help and encouragement; and to the Ordnance Survey for permission to reproduce the grid references.

Finally, the Editor thanks his daughter, Mrs J. R. Elliott, for her patience in producing the first typewritten script from a collection of preliminary and often much altered drafts.

W. J. Sivewright

Metric equivalents

Throughout this book Imperial measurements have been used in giving dimensions of the works described, as this system was used in the design of the great majority.

The following are metric equivalents of the Imperial units used.

Length
: 1 inch = 25.4 millimetres
1 foot = 0.3048 metre
1 yard = 0.9144 metre
1 mile = 1.609 kilometres

Area
: 1 square inch = 645.2 square millimetres
1 square foot = 0.0929 square metre
1 acre = 0.4047 hectare
1 square mile = 259 hectares

Volume
: 1 gallon = 4.546 litres
1 cubic yard = 0.7646 cubic metre

Flow rate
: 1000 gallons per hour = 1.263 litres per second
100 gallons per minute = 7.577 litres per second

Mass
: 1 pound = 0.4536 kilogram
1 UK ton = 1.016 tonnes

Pressure
: 1 pound force per square inch = 6.895 kilonewtons per square metre = 0.06895 bar

Power
: 1 horse power = 0.7457 kilowatt

Contents

Introduction

This book is based on the work of the Panel for Historical Engineering Works of the Institution of Civil Engineers, which was set up in 1971, and which has built up a considerable library of records of civil engineering works of interest.

The area covered by this volume continues southward from that of the volume on *Northern England* of the series *Civil Engineering Heritage*, and includes Wales and the western part of England to just south of Bristol.

It ranges over some of the wildest and some of the most industrialized parts of the country. At its heart is the Severn Valley around Coalbrookdale, where the former great iron industry founded by Abraham Darby I, once a Bristol brass founder, contributed largely to the rapid advance of the Industrial Revolution from the mid-eighteenth century. To-day the major industrial areas are around Birmingham, in South Wales and in the Bristol region, the last two served by major ports. Not to be forgotten are the salt and chemical industries of Cheshire and the Potteries of Staffordshire. It is not so long since slate quarrying was a staple industry in North Wales, while the Forest of Dean, much of whose timber went to building ships for the Royal Navy during the Napoleonic Wars, had important resources of coal, worked well into the present century. Both inland and along the coasts there are areas of considerable attraction to tourists and holiday makers.

To the civil engineer has fallen the task of providing and maintaining the infrastructure of industry and transport; of serving the needs of the population for water and drainage; and of protecting the environment against the ravages of natural forces.

The works described here have been chosen so as to cover as wide a field as possible and to give some idea of the scope and development of civil engineering, rather than merely to provide a gazetteer or list of statistics. The more notable achievements of eminent civil engineers such as Brindley, Telford, Jessop, Rennie and Brunel have of course been covered, but equally there are

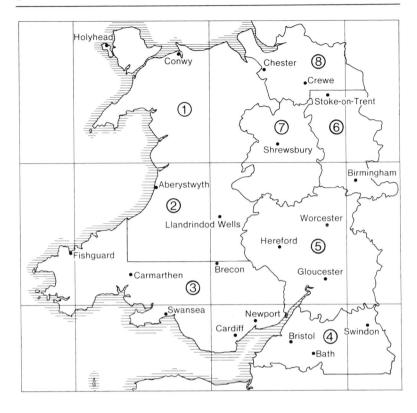

numbers of smaller or less well known works. All of them, however small, are either very attractive or have some special point of technical interest. It is hoped that not only civil engineers but also the general reader will be helped to understand what the civil engineering profession has done over the years for the economy, the environment and the convenience of man, and will derive some insight into the immense variety and range of expertise which have gone into our civil engineering heritage.

As in the volume on Northern England, the book is divided into chapters, each covering a particular area and each accompanied by a map and a list of the works described, together with a brief account of the background to civil engineering in the area.

The Ordnance Survey grid reference is given for each item, as is also the HEW (Historical Engineering Work) number under which the work is registered in the Library of the Institution of Civil Engineers.

The material for the book has been collated from much more fully detailed records prepared by Panel members in the various parts of the area. The book is by no means comprehensive and at the end is to be found a list of HEWs which have been recorded but which are not mentioned in the text. The work of the Panel is a continuing process and many further works are currently under investigation.

In addition to specific references, marked by superior numerals in the text, there is appended a bibliography of publications which deal broadly with the civil engineering history of the area or with particular types of structure.

CHAPTER 1

North Wales

The boundary between Wales and England commences in the estuary of the River Dee (Afon Dyfrdwy) and skirts Chester less than two miles from the city centre. The political boundary follows the Dee Valley, but the geographical boundary is the Clwydian Range, the backbone of the county of Clwyd. The Vale of Clwyd is the boundary with Gwynedd, a county famous for the natural grandeur of the Cambrian Mountains, which culminate in Snowdon, the highest mountain in either Wales or England. Though separated by the Menai Strait, Anglesey forms part of Gwynedd, still a stronghold of the Welsh language and culture.

The mountainous landscape of North Wales presents serious obstacles to communications and is complemented by many excellent examples of the science and craft of civil engineering, in particular the works of two of the most eminent engineers, Thomas Telford and Robert Stephenson. The route to Ireland via the harbour at Holyhead on Anglesey was of great political importance during the 18th and 19th centuries, and in constructing his highway through the mountains Telford built some of his finest works. Stephenson took an apparently easier coastal route for his railway, but this was not without its difficulties, and his ingenuity was pushed almost to its limits.

Slate quarrying was the most important industry in North Wales until the Second World War, and the steepness of the terrain dictated that narrow gauge rather than standard gauge railways were built to bring the slate down from the mountain quarries to the harbours, whence Welsh slate was exported all over the world. Nowadays the 'little trains' no longer haul slates, but are a considerable tourist attraction instead.

North Wales has provided the industrial heartland of England with drinking water and hydro-electricity for many years. The mountains just north of Snowdon form the backdrop for one of the most impressive schemes carried out in Britain during recent years – the hydro-electricity generating station at Dinorwig.

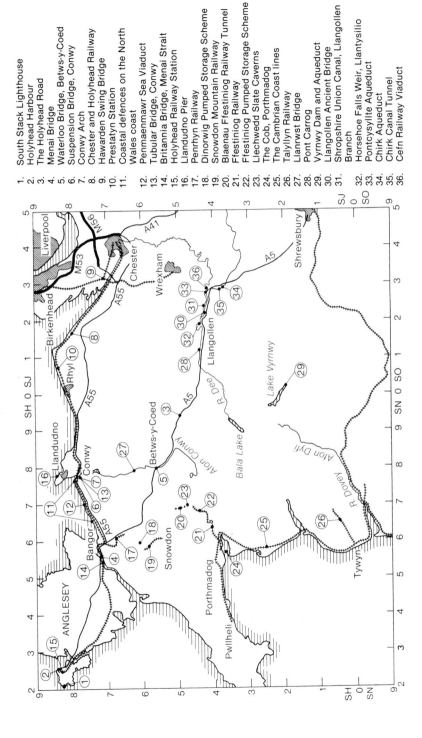

1. South Stack Lighthouse
2. Holyhead Harbour
3. The Holyhead Road
4. Menai Bridge
5. Waterloo Bridge, Betws-y-Coed
6. Suspension Bridge, Conwy
7. Conwy Arch
8. Chester and Holyhead Railway
9. Hawarden Swing Bridge
10. Prestatyn Station
11. Coastal defences on the North Wales coast
12. Penmaenmawr Sea Viaduct
13. Tubular Bridge, Conwy
14. Britannia Bridge, Menai Strait
15. Holyhead Railway Station
16. Llandudno Pier
17. Penrhyn Railway
18. Dinorwig Pumped Storage Scheme
19. Snowdon Mountain Railway
20. Blaenau Ffestiniog Railway Tunnel
21. Ffestiniog Railway
22. Ffestiniog Pumped Storage Scheme
23. Llechwedd Slate Caverns
24. The Cob, Porthmadog
25. The Cambrian Coast lines
26. Talyllyn Railway
27. Llanrwst Bridge
28. Pont Carrog
29. Vyrnwy Dam and Aqueduct
30. Llangollen Ancient Bridge
31. Shropshire Union Canal, Llangollen Branch
32. Horsehoe Falls Weir, Llantysilio
33. Pontcysyllte Aqueduct
34. Chirk Aqueduct
35. Chirk Canal Tunnel
36. Cefn Railway Viaduct

1. SOUTH STACK LIGHTHOUSE HEW 756
SH 202 823

The north shore of Anglesey is roughly on the same latitude as Liverpool, rather above the general line of the North Wales coast. The top left-hand corner, as it were, is therefore marked for the main shipping lanes by the Skerries Lighthouse.

Holyhead Harbour (HEW 1095) is on Holy Island, seven miles to the south, and is itself marked on the west side by the South Stack Lighthouse. In addition it has two local lights, the one on Admiralty Pier being 48 ft high and dating from 1821; the other 63 ft high on the outer breakwater, with both light and breakwater dating from 1873.

South Stack is a tiny rock island of high cliffs cut off by a hundred feet of 'turbulent sea'. Access to it is by 380 steps down to a slender aluminium truss bridge of 110 ft span, which in 1964 replaced a suspension footbridge of 1828. Before that there was only a breeches buoy arrangement.

South Stack Lighthouse was designed by Daniel Alexander and first lit on 9 February 1809. Built from stone quarried on the site, it consists of a tower of traditional form, tapered and painted white, with gallery and lantern about 90 ft high above the rock. The lighthouse is flanked by a long low building with a two-span pitched roof and three smaller buildings. These were the engine room and dwellings. The operation is now automatic.

Illumination was originally by oil, then by paraffin vapour and finally, from 1939, by electricity. The light is nearly 200 ft above mean high water and can be seen for over 20 miles.

2. HOLYHEAD HARBOUR HEW 1095
Illustrated on page 8 SH 250 840

The 63 miles route from Holyhead to Dun Laoghaire is the most direct sea crossing from Great Britain to Dublin.

The history of the port of Holyhead is complicated by the changing names of its elements.

In the 17th and 18th centuries there was just a creek, which dried out at low water, later known as the Inner Harbour. This was given protection by the Admiralty Pier (HEW 1097, SH 253 829), designed by John Rennie and completed in 1821, which ran eastward from Salt Island and was linked to the town by a swing

bridge over the Sound. The Custom House and Harbour Office on
Salt Island were part of the scheme. A triumphal arch com-
memorates the visit of George IV in 1821.

The next stage comprised Telford's Dry Dock of 1825 (HEW
1096, SH 254 826), opposite the tip of Admiralty Pier, and his
South Pier, 1831. The Admiralty Pier was then regarded as the
North Pier.

The Dry Dock was designed to drain at low water spring tides,
but by 1829 steam pumping was installed to empty it over the

Breakwater lighthouse, Holyhead

neaps. The Science Museum, London, has a photograph of the Boulton & Watt engine which drove the two bucket pumps. Original drawings suggest that the dock was built with hinged gates, but later a caisson was substituted. The dock is now derelict, although in 1939 it had a new caisson fitted for wartime purposes.

The New Harbour was authorized in 1847, the area within the North and South Piers becoming the Old Harbour.

A breakwater 7860 ft long (HEW 1099, SH 237 837 to SH 258 847), built with stone brought down by a 7 ft gauge railway (changed to standard gauge in 1913) from quarries on Holyhead Mountain, was under continuous construction from 1848 until 1873. The objective was to enclose a deep water sheltered roadstead of 400 acres, in addition to the 276 acres of harbour. The Engineer was J. M. Rendel, and on his death in 1856 the work was taken over by (Sir) John Hawkshaw, assisted by Harrison Hayter, who had already been engaged on the project for some years.[1,2]

The Chester and Holyhead Railway came in 1848, first with a terminus at the head of the creek, then in 1856 to the Admiralty Pier which, widened and extended in timber, became known as the Mail Pier, and was used by the Dublin mailboats from 1849 to 1925. The timber widening and extension were demolished between 1935 and 1942.

Between 1875 and 1880, the London and North Western Railway Company developed the Inner Harbour with two quay walls forming a V that accommodated the new passenger station (HEW 1098, SH 248 822). A new graving dock was built on the east side of the harbour. In 1922 the channel to the station berths was deepened by a rockbreaker dredger. Telford had used a diving bell to trim off submerged rock.

In the past two decades there have been a number of developments in the port, including a deep water wharf for importing aluminium ore, roll-on/roll-off berths on the Admiralty Pier and in the Inner Harbour, and a container berth in the latter.

---

3.   THE HOLYHEAD ROAD   HEW 1212
Illustrated on page 10   SJ 400 176 to SH 250 832

In the last years of the 18th and first years of the 19th centuries the road from London to Holyhead, for the sea crossing to Ireland for Members of Parliament and the Royal Mail, was second in importance only to that from London to Dover, but over much of its length was in a dreadful state of repair, nowhere more so than

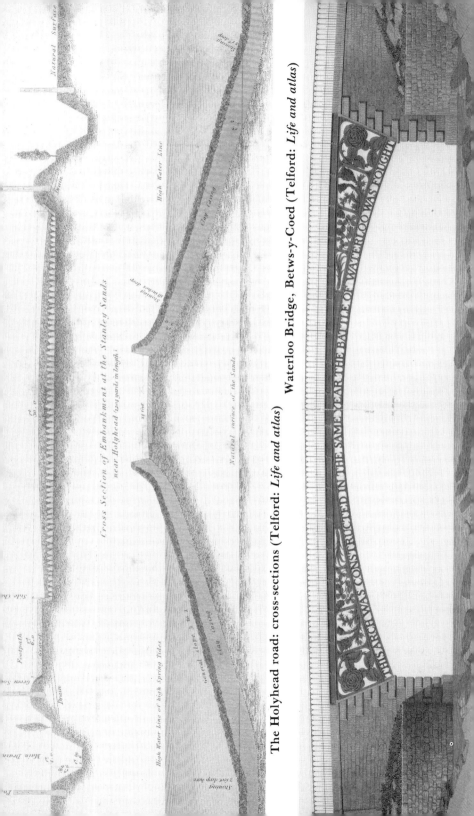

Cross Section of Embankment at the Stanley Sands
near Holyhead 1300 yards in length

The Holyhead road: cross-sections (Telford: *Life and atlas*)

THIS ARCH WAS CONSTRUCTED IN THE SAME YEAR THE BATTLE OF WATERLOO WAS FOUGHT

Waterloo Bridge, Betws-y-Coed (Telford: *Life and atlas*)

across North Wales. On Anglesey, reached by ferry, much of the route was only a grass track and through the Welsh mountains the road in places ran along the edges of unprotected precipices with gradients as steep as 1 in $6\frac{1}{2}$. Consequently travel was slow and dangerous; the Irish mail coach took 41 hours from London to Holyhead at an average speed of $6\frac{3}{4}$ miles per hour.

In 1810 Thomas Telford, who had been Surveyor of Shropshire since 1786, was commissioned to report on the state of the road and to suggest improvement, and in 1815 Parliament voted the necessary funds for what was one of the finest achievements by one of the greatest of British civil engineers.[3]

The Act required the route north-west from Shrewsbury, which had been under the administration of seven turnpike trusts, to be under the charge of one surveyor, who was to be a professional engineer, and Telford was appointed to the post.

By 1819 the 85 miles between Shrewsbury and Bangor had been made safe for traffic. Where the road passed through Glyn Duffrws, west of Corwen, the hillsides were blasted to enable it to be levelled and widened. From Rhydlanfair on the River Conwy, a three mile stretch was built along Dinas Hill with a maximum gradient of 1 in 22. After crossing the Conwy on the Waterloo Bridge (HEW 106) into Betws-y-Coed, the road followed the south bank of the River Llugwy to a point just upstream of the Swallow Falls, where it crossed the river and then took up a new alignment for one mile before rejoining the old line, and so on to Capel Curig. A new alignment through the Nant Ffrancon Pass (HEW 460, SH 722 582 to SH 626 659) led through Bethesda to Bangor.

Between 1820 and 1828 Telford built 20 miles of road across Anglesey from Menai Bridge (HEW 109) to Holyhead, including the Stanley Embankment (HEW 1246, SH 285 798 to SH 276 802) between Anglesey and Holy Island.

Telford's route is now the modern A5 trunk road. Although the turnpike system lapsed in the 1880s there remain here and there, as in other parts of the country, relics in the shape of tollhouses and roadmarkers. Among the former are those at Caergiliog (SH 304 786) and Llanfairpwllgwyngyll (HEW 459, SH 532 715).

---

4. MENAI BRIDGE　　　　　　　　　　HEW 109
Frontispiece　　　　　　　　　　　　　SH 556 715

In 1817 the Holyhead Road Commissioners instructed Telford to prepare plans for a bridge to replace the ferry across the Menai

Strait. The plans, for what is generally regarded as Telford's finest work,[4] were ready by February of the following year.

The construction took John Wilson, the contractor, a period of seven years from 1819. The bridge has an overall length of 1000 ft, seven stone approach spans of 52 ft, and a main central suspension span of 579 ft, tower to tower, carrying the road 100 ft above sea level.

The piers are faced with Anglesey marble. The deck was suspended from four sets of wrought iron chains.

Modifications to the bridge were made following damage by storm in January 1839.[5,6] In 1893 Sir Benjamin Baker replaced the timber deck by steel troughing on flat-bottomed rails. In 1940 the chains were replaced by two sets of steel chains, the deck was rebuilt in steel to take heavier road traffic and a cantilevered footway was added on each side, so giving the bridge the appearance that it has today.[7] Nevertheless, the modern alterations do not detract from the gracefulness of Telford's original structure, set as it is in magnificent scenery.[8]

A very full account of the design and construction is given by W. A. Provis, the resident engineer, in a large folio volume published in 1828.[9]

5.   WATERLOO BRIDGE, BETWS-Y-COED   HEW 106
Illustrated on page 11   SH 798 557

This well known bridge was built by Telford in 1815 to carry the London to Holyhead road over the Afon Conwy. Its single span of 105 ft consists of five cast iron arched girders at 5 ft centres supporting cast iron deck plates. The outer ribs carry the legend 'This arch was constructed in the same year the battle of Waterloo was fought', and the spandrels above are beautifully decorated with rose, thistle, shamrock and leek, modelled in relief by William Hazledine, whose foreman, William Stuttle, was responsible for the erection.

In 1923 the bridge was strengthened by concreting the inner three ribs and adding a 7 in. reinforced concrete deck. This was cantilevered to provide new footpaths and allow the roadway to be widened.

In 1978 a new 10 in. reinforced concrete deck was added and the original cast iron parapet fence, visible from outside, was protected by an additional fence inside. The masonry abutments were strengthened on both occasions.

6. SUSPENSION BRIDGE, CONWY   HEW 107
Illustrated on page 14   SH 785 776

Opened in 1826, this was the most successful of Thomas Telford's six 'Gothic' bridges.[10] Two pairs of solid ashlar limestone towers 12 ft 4 in. in dia. and 40 ft high support the chains, which span 327 ft between their supports on the towers. They are linked by castellated walls containing the 10 ft wide carriageway arches. The wrought iron chains are arranged in two tiers of five 9 ft long links joined by deeper plates, the joints of which are alternated. Vertical rods at 5 ft intervals carry the deck suspended from the junction plates. The present deck is a replacement dating from 1896. The original deck was probably made up of a light iron framework braced by bars on its underside, upon which were laid two longitudinal layers of fir planks. Plans to demolish the bridge following the construction of a new road bridge (HEW 692) alongside led to a world outcry in 1958, since when the bridge has been in the care of the National Trust and closed to vehicular traffic.

7. CONWY ARCH   HEW 692
Illustrated on page 14   SH 785 775

The 310 ft span steel arch at Conwy, built in 1958 to relieve Telford's suspension bridge (HEW 107) of the weight of road traffic, is a very fair solution to the problem of blending a modern structure into a pleasing environment, which in this case is dominated by Conwy Castle and by the Telford and Stephenson (HEW 108) bridges immediately upstream.[11]

It was intended to build a bridge twice as wide and this is only the first half. The two sides differ. The upstream side shows a simple spandrel-braced arch of N type and a short approach viaduct of short spans on columns.

The castle and bridges are mostly viewed from downstream, from which direction the viaduct is masked by what appears to be a long wing wall with a pilaster resembling an old type cutwater and proportioned like a turret of the castle. From this springs a flat and graceful arch which, combined with the vertical flat curve of the roadway, gives the general impression of an ancient road bridge. The castellated towers of the other two bridges can just be seen in the background, and the simplicity of the newer bridge makes a satisfactory contrast enhancing both, as can be seen from the coloured tourist postcards.

The new A55 road crossing of the river further downstream is to be hidden from view by being carried in immersed tubes made in reinforced concrete cast on land, floated out and sunk in position.

8.	**THE CHESTER AND**	**HEW 1094**
	HOLYHEAD RAILWAY	SJ 413 670 to SH 248 822

The success of the Liverpool and Manchester Railway (HEW 223)[12] showed that rail travel offered speeds three times those possible on the roads, and effectively made Telford's excellent improvements to the Holyhead Road (HEW 1212, HEW 1213, 1214) out of date almost as soon as they were completed. Communications between England and Ireland remained important. Of three railway schemes which were considered for this area, the one which was adopted was that of the Chester and Holyhead Railway, incorporated in 1844 with Robert Stephenson as Engineer-in-Chief and completed in 1850 over a route surveyed by his father in 1838. Hedworth Lee was Resident Engineer.[13]

Conwy bridges: tubular bridge, suspension bridge and steel arch bridge (Aerofilms Ltd)

The line skirts the North Wales coast, and for some 43 miles the special feature of the route was the need for heavy engineering works to protect the railway and the coast from the sea (HEW 1228, 829). The new A55 road along the coast faced similar problems, which modern engineers have dealt with by adopting solutions similar to those of Stephenson.

Important viaducts on the railway are the 22-arch structure at Talybont (HEW 1280, SH 602 707), just north of which is Penrhyn Castle, and that at Malltraeth (HEW 1288, SH 414 691) on Anglesey, but of prime importance are the river crossing at Conwy (HEW 108) and the Britannia Bridge over the Menai Strait (HEW 110).

Near its western terminus, the railway crosses the strait between Anglesey and Holy Island alongside Telford's Stanley Embankment (HEW 1246).

Francis Thompson designed the original stations as two-storey rectangular structures in the Georgian style. Of these one of the smallest, which is the first reached after crossing the Menai Strait, is the best known to tourists, solely because of the length of its name Llanfairpwllgwyngyllgogerychwyrndrobwllllantysiliogogogoch, popularly abbreviated to Llanfair P.G.

The Chester and Holyhead railway was built for speed, a feature emphasized in 1857 when John Ramsbottom installed the first ever water troughs near Mochdre, later resited near Aber (SH 652 724). The troughs were several hundred yards long and located between the rails. By lowering scoops into the troughs, the locomotives picked up water and this made non-stop long distance rail travel a reality by avoiding the necessity for trains to stop for rewatering.

---

9.   HAWARDEN SWING BRIDGE,   HEW 775
SHOTTON   SJ 312 694

Three spans of steel hogback N trusses carry two railway tracks over the River Dee. The two fixed spans are each 125 ft and the swing section is of record length 287 ft, which gave a clear opening of 140 ft. Movement was originally by hydraulic rams and chains to a 32 ft dia. circular girder. The 90 ft landward end of the bridge ran on a quadrant rail. The bridge was fixed and the machinery removed in 1971. The girders have a maximum depth of 32 ft and are at 27 ft 6 in. centres. There are cross girders at 17 ft centres with two pairs of rail girder bearers.

The bridge was built in 1887 – 89 by John Cochrane & Sons for the Manchester, Sheffield and Lincolnshire Railway (later part of the Great Central), which in spite of its name had a line as far into Wales as Wrexham.

Francis Fox[14] was the engineer and the bridge was opened on 3 August 1889 by the wife of W. E. Gladstone, after whose residence it was named.[15]

---

10.   PRESTATYN STATION   HEW 1208
                                                          SJ 064 831

Prestatyn Station, on the line of the Chester and Holyhead Railway (HEW 1094), is one of the few remaining examples of the prefabricated single-storey station buildings manufactured at the Crewe Works of the London and North Western Railway in the closing years of the 19th century.

The buildings were constructed in 6 ft 8 in. modules and the timber framework was faced externally with rusticated boarding. Brick footings, fireplaces and chimneys were used where required. Canopies with valances were cantilevered over the platforms, their timber beams being supported on iron brackets. Prestatyn Station was rebuilt in the original style in 1979.

---

11.   COASTAL DEFENCES ON   HEW 1228
         THE NORTH WALES COAST   SJ 20 78 to SH 68 76

About half of the entire length of the Chester and Holyhead Railway (HEW 1094) is beset with unending maintenance problems arising from river and tidal flooding, coastal erosion and storm damage, and numerous important engineering works have had to be built.

At Holywell $2\frac{1}{2}$ miles of embankment have constantly to be replenished by tipping slag some 100 yd from the railway. There is another wall 3900 yd long between Mostyn and Point of Ayr.

The Abergele railway embankment, known as 'The Cob', consists of 1960 yd of shingle beach and groynes flanked by the 1470 yd Rhuddlan Marsh wall, first embanked in 1800 by Trustees and raised and improved when the L&NWR took it over in 1880.

At Llysfaen, east of Penmaenrhos Tunnel, landslips have frequently occurred, only partly compensated until now by their having produced material to replenish the beaches to the east. Currently the problem here is in effect being transferred from rail to road. The new A55 dual carriageway[16,17] is being built to seaward of the railway and is being protected by some 22 000 'dolosse', specially shaped interlocking concrete objects, weighing five tons each and placed on a 1 in 2 slope as facing to secondary armour of one ton rockfill, with a smaller rock core and a filter membrane.

At Old Colwyn there is an extensive area of unstable ground below the railway. For 680 yd there are groynes, toe walls, larger walls, pitched slopes and wave-breakers.

From Colwyn Bay to Conwy, protection is given by the promontory that terminates in the Great Ormes Head.

Further west, as far as Llanfairfechan, the problem is different. High cliffs forced the coach roads of Telford's day and earlier and then Stephenson's railway, to make use of shelves, either on sea walls at the base or higher up on the cliffs themselves, and occasionally to burrow through headlands in tunnels.

East of Penmaenmawr half a mile of railway is protected by a 20 ft wall leading to the Penmaenbach tunnel. Part of this wall was destroyed in 1945 and was replaced to a new design in reinforced concrete.

Penmaenmawr Sea Viaduct (British Rail)

The 253 yd Penmaenmawr tunnel is extended by avalanche shelters to guard against rockfalls from the cliffs towering above it.

The Penmaenmawr Sea Viaduct (HEW 829) is part of a 1500 yd length of solid masonry sea wall up to 40 ft high, located some 20 ft above High Water Level and surmounting a 1 in 30 pitched slope down to the beach.

Coastal defence is an important branch of civil engineering, often dramatic and always expensive. By no means, of course, is it confined to railways, although other examples of such use are to be found between Dawlish and Teignmouth in Devon, and in Kent between Folkestone and Dover.

---

12.   PENMAENMAWR SEA VIADUCT   HEW 829
Illustrated on page 17   SH 696 762

On 22 October 1846, during the construction of the sea wall at Penmaenmawr to protect the Chester and Holyhead Railway (HEW 1094, 1228), a severe storm with 40 ft waves destroyed the

Britannia Bridge, Menai Straits: Robert Stephenson is seated in centre and Joseph Locke and I. K. Brunel are seated at right (Lucas, The Institution of Civil Engineers)

most exposed section of the wall in the presence of the engineer, Robert Stephenson.

He decided to replace it by an open viaduct, some 182 yd long, of 13 equal spans through which the waves could dissipate their energy up the sloping beach beneath.

The masonry piers, which were built in cofferdams, are 32 ft wide and 6 ft 4 in. thick and stand to a height of 41 ft, about 15 ft above high water. They were protected from scour by piles retaining large boulders set in concrete to form a pavement extending some 18 ft seaward.

The original deck consisted of four longitudinal cast iron girders under each track, of inverted T-section, 24 in. deep, 14 in. wide at the bottom and $4\frac{1}{2}$ in. at the top, carrying timber way beams. The parapet girders were similar. The decking was of timber laid crosswise.

Work began in March 1847 and the viaduct opened to traffic on 1 May 1848.

In 1908 the deck system was replaced by 6-ring brick arches of 13 ft rise.

---

13. TUBULAR BRIDGE, CONWY	**HEW 108**
Illustrated on page 14	SH 785 774

This bridge was built in 1848 by Robert Stephenson to carry the Chester and Holyhead Railway across the river Conwy. It is very similar in design to his other great bridge on the same line, the Britannia Bridge (HEW 110) and like it, marked a significant advance in the art of bridge building.[18] It consists of a single span 400 ft long, formed by two parallel rectangular wrought iron tubes, each weighing 1300 tons. These were built ashore and then floated out on pontoons to be raised into position onto stone abutments on either side of the river. Masonry towers were built on the abutments and topped with battlements and turrets to harmonize with the adjacent Conwy Castle.

Before the bridge was commissioned, loading tests were carried out on both the tubes using up to 300 tons of iron kentledge, which produced a central deflection of 3 in. By way of comparison, the passage of an ordinary train is said to produce a deflection of only $\frac{1}{8}$ in.

Two piers added in 1899 reduced the span by 90 ft.

The railway approaches the bridge beside the embankment that Telford built to his elegant suspension bridge (HEW 107).

14. BRITANNIA BRIDGE, MENAI STRAIT   HEW 110
Illustrated on page 18   SH 542 710

Opened in 1850 to carry the Chester and Holyhead Railway across the Menai Strait, the Britannia tubular bridge, together with that at Conwy (HEW 108), may be said to be the forerunner of the box girders of today.

Robert Stephenson contemplated using suspension spans with deep rectangular trough-shaped stiffening girders, but was struck with the idea that if the top of the trough were closed in, the girder might be self-supporting. Studies by Professor Eaton Hodgkinson into the strength of materials, and careful testing of large models by the structural engineer and shipbuilder William Fairbairn confirmed the feasibility of this approach.[19]

As at Conwy, the tracks were carried within twin rectangular rivetted tubes built up from wrought iron plates, as developed for shipbuilding. The two main spans were each of 460 ft, flanked by side spans of 230 ft. Hitherto the longest wrought iron span had been 31 ft 6 in., so the successful building of the Britannia and Conwy bridges was an outstanding advance in the use of this material in structures.

The four tubes for the two mainstream spans, each weighing 1800 tons, were built on the Caernarfon shore, then floated out and jacked up 100 ft onto the towers, a procedure like that at Conwy (HEW 108) and subsequently adopted for his Chepstow and Saltash bridges by I. K. Brunel. In June 1849, Brunel stood beside his great friend and rival Stephenson to watch the launching of the first Menai tube.[20]

All four spans for each track were connected end to end through the towers to form a 1511 ft long continuous girder and so take advantage of the economy of material stemming from continuity.

With its adherence to straight lines, its massive masonry abutments and its three masonry towers rising over 100 ft above the bridge, the structure presented an 'outstanding example of simplicity, symmetry, harmony and proportion'.[21]

In 1970 a fire destroyed the protective timber roof above the tubes, and the heat caused the tubes to tear apart into four separate spans, so losing continuity and strength.

The bridge was replaced by a clever, practical and modern structure of quite different design,[22] which still retains Stephenson's original stone towers complete with the monumental lions at the ends of the bridge. The new main spans are steel arches with eight panels of N-truss spandrel bracing in each half arch, an arrangement very similar to that of the longer Victoria Bridge built some 70 years earlier over the Zambesi River between Zimbabwe and Zambia.[23] Each side span was divided into three spans built in reinforced concrete. Both bridges were built by the Cleveland Bridge and Engineering Co. of Darlington. The new Britannia bridge carries above the rail tracks a road onto which the A5 trunk road was diverted, so that Telford's Menai Bridge (HEW 109) was relieved of heavy traffic.

---

15.   HOLYHEAD RAILWAY STATION   HEW 1098
SH 248 822

The development of Holyhead Inner Harbour by the London and North Western Railway included a large new station and hotel, which were opened by HRH the Prince of Wales in June 1880.

The layout, which must be unique, was a V, with the down side tracks and buildings along the west arm of the Harbour, and the up along the east arm, with the hotel at the south end in the V. Only the up side remains. It is a good example of the type of overall roof used by the L&NWR at that time for several of their principal stations, as at Huddersfield (HEW 240, SE 145 168).[24]

The site is curved and tapering and the 64 trusses vary in span from 62 ft downwards. They are supported on a substantial screen wall 25 ft high on the one side, and on cast iron columns with short wrought iron lattice girders between, on the other.

With a 1 in 2 roof slope and a camber of about a foot in the main tie, the trusses are of distinctive design, with a clerestory carrying louvred ventilators. The main queen posts are braced by four tie rods and a ring.

---

16.   LLANDUDNO PIER   HEW 432
SH 784 830

Llandudno Pier was designed by James Brunlees and Alexander McKerrow and built by John Dixon in 1876. The 2295 ft long structure is in two sections. The main pier is carried on wrought iron lattice girder framework supported on cast iron columns, and extends 1234 ft to a T-shaped pier head 60 ft wide. The deck is lined with four pairs of kiosks, with three larger kiosks at the head. At the shore end, an arm of the platform connects with the pavilion some

distance along the shore. The pavilion has a projecting gable centre portion, and recessed wings with apsidal ends, and is fronted by
a verandah.

---

17. PENRHYN RAILWAY   HEW 1257
SH 592 730 to SH 616 660

The Cegin viaduct on the Chester and Holyhead Railway crosses
the line of the narrow gauge Penrhyn Railway which used to carry
slate to Port Penrhyn from quarries at Cefn-y-Parc near Bethesda
some 6½ miles inland. These were in use as early as 1580 and were
served by packhorses. In 1782 large scale working began and the
roads were improved. In 1801 packhorses were superseded by a
horse waggonway, one of the few in Wales to use edge rails and
flanged wheels. The first rails were unusual, being of oval section.
The line was rebuilt to 1 ft 10¾ in. gauge in 1874 with steam
locomotives, as many as 28 being at work at one time, mostly in the
quarry itself, which is now served by road. From 1879 to 1958 the
line carried workmen as passengers. It finally closed in 1962.

An excellent selection of relics from the railway, of which a few
traces remain on the ground, is preserved in the Industrial
Railway Museum opened in 1965 in the stable block of Penrhyn
Castle (SH 603 720).[25] Some of the permanent way switch and
crossing items are particularly interesting.

One of the narrow gauge locos on display is the 0-4-0T Hunslet
'Charles' of 1882, whose sisters 'Linda' and 'Blanche' are in use
on the Ffestiniog Railway now. An unusual locomotive to be
preserved is the standard gauge ex-L&NWR 4 ft 3 in. Webb
0-6-2T built at Crewe in 1888.

Penrhyn Castle itself was designed by Thomas Hopper for Mr
G. H. D. Pennant and built between 1827 and 1837. It now
belongs to the National Trust.

---

18. DINORWIG PUMPED STORAGE   HEW 1236
SCHEME   SH 598 607
Illustrated on facing page

On the opposite side of Llyn Peris from the lower terminus of the
Snowdon Mountain Railway (HEW 1222) is the site of the largest
pumped storage power station in Europe and the third largest in
the world.[26]

As the name implies, pumped storage is in effect a means of storing energy. Electricity generated from base-load stations at periods of low demand is used to pump water from one reservoir to another at a higher level. At periods of peak demand, water is released to the lower reservoir passing on its way through turbines which drive electric generators feeding to the national grid. This avoids the necessity for providing expensive thermal stations solely to deal with peak loads or for operating base-load stations at below their economic capacity in order that output can be increased to meet fluctuations in demand. Feeding the pumping mode of the pumped storage system permits more constant and therefore more economic operation of thermal base-load stations. Hydro-electric stations can be brought on power in seconds, which makes this system invaluable for dealing with sudden demands such as occur at the ends of popular television programmes or on the occasion of unexpected breakdowns on the grid.

The upper reservoir at Dinorwig is the existing lake of Marchlyn Mawr whose capacity has been raised by a 1970 ft long rockfill

Dinorwig Pumped Storage Scheme: power station machine hall (Central Electricity Generating Board)

dam 118 ft high (HEW 1237, SH 62 63), which permits the water to rise 108 ft above its original level to accommodate the operational fluctuation of 100 ft. The upstream face of the dam is sealed by a layer of asphaltic concrete, only the second instance of such an application in Britain.

From Marchlyn Mawr 5575 ft of $34\frac{1}{2}$ ft dia. low pressure tunnel lead on a slight gradient to a 33 ft dia. vertical shaft 1475 ft deep. From the foot of this, a 31 ft dia. high pressure tunnel leads to the power station about 2200 ft away, dividing into six smaller diameter tunnels before reaching the turbines.

The machinery is housed in nine man-made caverns under the Elidir Mountain, the largest, the main machinery hall, being 590 ft long, 80 ft wide and 197 high. The nearby transformer hall is 530 ft long, 80 ft wide and 62 ft high; there are other massive shafts and galleries for hydraulic and electric control equipment.

The six turbines work in reverse as pumps with their alternators as motors. On average, pumping lasts six hours per night, and generation five hours per day, with an average output of 1680 MW at 18 000 V from an installed capacity of 1800 MW. A load up to 1320 MW can be picked up in ten seconds. From the turbines, six 12 ft dia. tunnels lead in pairs to three 27 ft dia. tail-race tunnels which discharge into Llyn Peris, some 2000 ft away and 1650 ft below Marchlyn Mawr.

The project was designed by the Central Electricity Generating Board with James Williamson of Glasgow, in association with Binnie and Partners of London, as consultants. When the main underground works were let in November 1975, a record was set for the highest value civil engineering contract ever let in the United Kingdom.

The station was fully commissioned at the end of 1983 and formally opened by HRH the Prince of Wales on 9 May 1984. The total capital expenditure of £425 000 000 is expected to save the CEGB £50 000 000 per annum.

Great care was taken to minimize the effect of the works on the environment, in the heart of Snowdonia. Practically the whole of the construction is under ground, even the 400 kV outgoing transmission cables being buried for a distance of six miles. The few administrative buildings above ground are of local stone, much of it from old quarry buildings. A good deal of the spoil from the excavations, which include a total of ten miles of tunnels, together with that from the removal of heaps of slate waste which had disfigured the neighbourhood for years, was tipped into old quarries or the deeper parts of Llyn Peris.

Snowdon Mountain Railway: first day of operation

A short distance away at Llanberis is the water wheel (HEW 1284, SH 585 602), one of the largest in the United Kingdom, which drove the machinery for the Dinorwig slate works, now the Welsh Slate Museum, part of the National Museum of Wales.

19. THE SNOWDON MOUNTAIN RAILWAY HEW 1222
Illustrated above   SH 582 597 to SH 609 543

The Snowdon Mountain Railway has the double distinction of being the only rack railway in Britain, and having the highest station, at 3493 ft above sea level. It was designed by Douglas and Francis Fox, using the Swiss system developed by Dr Roman Abt, and was built by A. H. Holme and C. W. King of Liverpool between December 1894 and January 1896, their men working a five day week, unusual in those days, because of the arduous conditions on the mountainside. The line rises 3140 ft in a distance of 4 miles 1100 yd, from Llanberis to the Summit Station on Yr Wyddfa, the highest peak on Snowdon, at an average gradient of 1 in 7.8, the steepest being 1 in 5.5, and the mildest 1 in 50. The line was designed to the European narrow gauge of 800 mm (2 ft 7½ in.) for the running rails, which are attached to steel sleepers. The two central rack blades are bolted either side of steel chairs, which are in turn bolted centrally to each sleeper. The line is single track, with passing loops at the stations.

There are ten bridges, the largest being the Lower Viaduct over Afon Hwch with 13 arches spanning 30 ft and one skew arch spanning 38 ft 4 in. over a road. This structure is 500 ft long and is built on a gradient of 1 in 8.5.

20.	BLAENAU FFESTINIOG	HEW 755
	RAILWAY TUNNEL	SH 688 503 to SH 697 469

This 2 mile 340 yd long tunnel under 1712 ft high Moel Drynogydd is the seventh longest railway tunnel in Britain.[27] It was designed initially by Hedworth Lee of the London and North Western Railway to take a narrow-gauge single track line from Betws-y-Coed to Blaenau Ffestiniog to join up with the Ffestiniog Railway,[28] and construction was commenced on 6 December 1873 with the contractor Gethin Jones starting work on the northern of the three shafts used for the excavation of the tunnel. Eight headings were used from the bases of the shafts and the portals, and extremely hard stone was encountered, causing Jones to abandon the work, the tunnel being completed by direct labour under the supervision of William Smith, who succeeded Lee as District Engineer at Bangor.

During construction, the LNWR decided to alter the line to standard gauge, although this meant dispensing with the connection to the Ffestiniog Railway, and altering the work already done. The tunnel is 18 ft 6 in. high and 16 ft 6 in. wide and is straight except for short curved sections at either end. Most of the tunnel is on an ascending or descending gradient of 1 in 660 except for a short level at the summit near the southern end of the tunnel where the rail level is 673 ft above that at Betws-y-Coed Station. The tunnel was opened to traffic on 22 July 1879.

21.	FFESTINIOG RAILWAY	HEW 647
		SH 571 384 to SH 697 461

This 1 ft $11\frac{1}{2}$ in. gauge railway was designed by James Spooner and two of his sons and was opened in 1836 to take slate from the Blaenau Ffestiniog quarries to the harbour at Porthmadog.[29,30] It descended 700 ft in 13 miles, the maximum gradient being 1 in 80. Upgoing trains were pulled by horses for most of its length, except on inclines which were redesigned by Robert Stephenson with waterwheels to provide the traction. Downhill the horses rode in a special car with the loaded wagons going down under gravity.

The line crosses the bay at Porthmadog on the Cob (HEW 1192). The carefully graded route necessitated embankments up to 60 ft high, built of slate with almost vertical sides, and two tunnels at Garnedd and Moelwyn. Just below Tan-y-Bwlch station, the line crosses the B4410 at SH 647 414 on a neat little cast iron arch bridge of 18 ft skew and 12 ft square span (HEW 257, SH 648 475), built in 1854.

Steam was introduced in 1863 by James Spooner's son Charles, and passenger traffic began in 1864. The line was closed in 1946 but restoration work began by the Ffestiniog Railway Society, many of whose members rebuilt the railway in their own time to make it a highly successful tourist attraction.[31] The construction of the Ffestiniog power station (HEW 1238) meant diverting part of the line. The diversion begins at SH 679 422 just beyond Dduallt Station by turning off to the east and then making a complete loop clockwise (the only railway spiral in Britain), passing over the original line just south of the station, before turning north again to pass through a new tunnel at Moelwyn, which was opened in 1977, and thence along the west side of the Tan-y-Grisiau reservoir.

The line was reopened throughout its length in May 1982.

---

22. FFESTINIOG PUMPED STORAGE SCHEME

HEW 1238
SH 679 449

This was the first pumped storage power station in Britain.[32] The upper reservoir was formed by enlarging Llyn Stwlan by a concrete dam 1250 ft long and 110 ft high, the lower, 1033 ft below, by damming the Afon Ystradau near the village of Tan-y-Grisiau. Here the concrete dam is 1855 ft long and 50 ft high. From the upper reservoir two vertical shafts fall 640 ft to two pressure tunnels which slope towards the power station for 3750 ft and then feed four steel pipes leading to the turbines and pumps a further 700 ft away.

The four alternators, each of 90 MW output when driven by the turbines, can operate as motors to drive the pumps. The latter are uncoupled when the station is generating. For pumping, the pumps are started by the turbines and when synchronization of frequency with the system has been reached, the alternators become driving motors and the water to the turbines is cut off.

The power station building on the west side of the Tan-y-Grisiau reservoir is of steel-framed construction faced in local stone. At 237 ft long, 72 ft wide and 66 ft high above ground, it is probably the largest stone building in Wales built since Harlech and Cricieth castles.

As at Dinorwig, care was taken with landscaping. The spoil from Stwlan dam was placed in the reservoir and the face of the Tan-y-Grisiau dam is concealed from view by rock from the excavations.

General design was by the Central Electricity Generating Board with Freeman Fox and Partners, in association with James Williamson and Partners, and Kennedy and Donkin, as consultants. The main contractors for the civil engineering work were the Cementation Co. Ltd and Sir Alfred McAlpine and Son Ltd. Work began early in 1957 and was completed in March 1963.

---

23.   LLECHWEDD SLATE CAVERNS   HEW 1169
SH 698 475

The Ordovician slate at Llechwedd lies in five beds inclined at 30 degrees to the horizontal, and is sandwiched between layers of

Barmouth Viaduct (British Rail)

hard chert. It was discovered in 1849 by John W. Greaves. It was mined by boring and shot-firing using gunpowder to form alternating chambers and pillars, which were carefully aligned vertically through over 1000 ft of rock to ensure support for all 16 levels in the mine.

There are five levels above the present-day tourist level, and ten below, though true mining is only carried out on the second level below the tourist level, modern quarrying techniques being used from the surface to remove the pillars in the uppermost four former levels. Dewatering used to be by surface water power, later replaced by electricity, but nowadays the mines are allowed to drain through an adit at a lower level. There is a museum containing some interesting slate processing machinery devised by Greaves and originally powered from an overhead shaft driven by an overshot water wheel. The slate was taken to the dock at Porthmadog on the Ffestiniog Railway, to connect with which Greaves installed a railway incline at Llechwedd in 1854.

---

24.   THE COB, PORTHMADOG   HEW 1192
SH 572 384 to SH 584 378

The construction of the embankment known locally as Y Cob was sponsored by William Alexander Madocks, MP for Boston, to reclaim 7000 acres of land from the tidal waters of Afon Glaslyn about five miles before the river enters Tremadog Bay. The rock-fill embankment, 90 ft wide at its base, 18 ft wide at the top and 21 ft high, was built on rush matting over a length of approximately 1400 yd between Boston Lodge on a hilly peninsula called Penryhn-isaf at its eastern end and the harbour of Porthmadog at its western end. It was built by John Williams between 1808 and 1811 and carries what is now the A487 road between Porthmadog and Maentwrog, the toll from motorists being collected at the eastern end.

The Ffestiniog Railway (HEW 647) also makes use of the embankment on the southern side of the road.

---

25.   THE CAMBRIAN COAST LINES   HEW 1227
Illustrated on facing page   SH 37 35 to SN 69 98 and SN 58 82

The Cambrian Coast lines round Cardigan Bay connect Pwllheli, Cricieth, Porthmadog, Harlech, Barmouth, Tywyn and

Aberystwyth to the rest of British Rail via Dovey Junction and Machynlleth. They also form a standard gauge link between the now active narrow gauge lines – the Great Little Railways of Wales – such as the Ffestiniog (HEW 647), Talyllyn (HEW 1210) and Vale of Rheidol (HEW 1121, Chapter 2).

Authorized in 1861–62, the Cambrian lines were opened in 1863–67. They extend for 55 miles from Dovey Junction to Pwllheli, and about 15 miles to Aberystwyth. The railway skirts the sea briefly at Pwllheli, Afon Wen, Cricieth and Harlech; then from Llanaber, north of Barmouth, to Tywyn, almost continuously. Further stretches occur on both sides of the Dovey Estuary, making a total of about 30 miles subject to coastal hazards.

The most difficult sections to maintain are at Llanaber (SH 5918) north of Barmouth, where huge concrete blocks connected by chains form part of the railway protective works, and at Friog, near Fairbourne (SH 6213) where the line is carried on a cliff shelf, and protected from rockfalls by an avalanche shelter. This has a 12 in. reinforced concrete roof slab some 60 yd long supported on reinforced concrete arches.

Crossing the Mawddach Estuary south of Barmouth at SH 623 151 is a remarkable viaduct (HEW 1167).[33] A rock foundation exists only at the northern end where there are two fixed spans of 37 ft 9 in.

The navigable channel, which is near the north shore, was originally crossed by an unusual opening span which tilted and drew back over the track. Now there is a swing span of 136 ft with a central pivot and a fixed span of 118 ft, both hogback trusses carried on cylindrical piers which replaced the original cast iron screw piles. To the south of the channel there are 113 openings of 18 ft span on timber pile trestles which are being replaced, or encased in glass fibre reinforced concrete sleeves as a protection against attack by marine borer.[34]

---

26.   TALYLLYN RAILWAY   **HEW 1210**
                                            SH 586 005 to SH 671 064

In the 1830s, slate began to be quarried at Bryn Eglwys, a remote site above the village of Abergynolwyn about $7\frac{1}{2}$ miles north-east of Tywyn on the Welsh coast. Packhorses took the finished slates to Aberdovey, but following the purchase of the quarries by William McConnel, a railway line was planned from Bryn Eglwys to Aberdovey. After the opening of the Aberystwyth and Welsh Coast

Railway between Aberdovey and Llwyngwril in 1863, the terminus of the line was changed to Tywyn to enable traffic to be interchanged with the new coastal railway.

The Engineer of the line was James Swinton Spooner, whose father and brothers were associated with the construction and development of the Ffestiniog Railway (HEW 647). Built to a gauge of 2 ft 3 in. (the minimum allowed by the Act), the line runs almost straight from Tywyn (Wharf) station, where there is a narrow-gauge museum, to Abergynolwyn station, a distance of just over $6\frac{1}{2}$ miles, climbing almost continuously along the south side of the valley of the Afon Fathew. Public passenger services terminated at Abergynolwyn but the line continued for a further $\frac{3}{4}$ mile to the foot of the first of several inclines connecting with the quarry workings. The maximum gradient of the line is 1 in 60. There were several intermediate stations, the most important being at Tywyn (Pendre), where the workshops and engine sheds were situated and where the passenger services from Abergynolwyn terminated. Just west of Dolgoch station, the line crosses the Nant Dolgoch at SH 650 045 on a three-span brick viaduct about 50 ft above the stream, the line's most impressive feature. The railway was opened in December 1866.

By 1911 the controlling interest in the railway had passed from the McConnel family to (Sir) Henry Haydn Jones MP. The slate quarries finally closed in 1946 and after the death of the owner in 1950 the future of the line was in jeopardy, but a Preservation Society was formed and took over the running of the line, which now forms a major tourist attraction. For a time L. T. C. Rolt, the author of several books on engineering history, acted as General Manager. In 1976 an extension of the public passenger service beyond Abergynolwyn to a new terminus at Nant Gwernol was opened.

27.   LLANRWST BRIDGE   HEW 164
                                                                           SH 798 615

This well-proportioned three-arch bridge carries a road to Gwydir Castle over the Afon Conwy in a most attractive setting. It is known as Pont Fawr, the 'Great Bridge' or the 'Shaking Bridge', as it vibrates if the parapet is struck at a point above the central arch, which has a span of 60 ft. The side spans are 45 ft and the bridge was built in 1636, reputedly by Inigo Jones. The bridge has a total length of 169 ft and the width between the masonry parapets is 13 ft 2 in. They carry on their outsides, at the centre, coats of arms and the date of construction.

Centrally placed on top of one parapet is a bronze sundial erected to mark the tercentenary of the bridge, unfortunately at a point where it is inadvisable for pedestrians to linger owing to the traffic and the narrowness of the bridge. The construction of the segmental arch rings is unusual. Although the lower voussoir rings are of normal shape with a slight taper inwards, the 4 in. by 20 in. slabs that form the second ring are dressed and laid to the curve, with their long concave faces downwards on top of the lower voussoir ring.

---

28. PONT CARROG	**HEW 682**
	SJ 115 437

On the Upper Dee is Pont Carrog which was probably built in the 17th century. It is constructed of local stone and carries the B5436. It has five segmental arches and the cutwaters extend up to the parapet walls to form recesses at road level. Its total length is 186 ft and the roadway extends the full width of 12 ft between the parapet walls.

---

29. VYRNWY DAM	**HEW 214, SJ 01 20**
VYRNWY AQUEDUCT	**HEW 1147**
Illustrated on facing page	**SJ 01 20 to SJ 47 93**

The building of the Vyrnwy Dam to supply water to the City of Liverpool marked the introduction to Britain of the high masonry dam. Work began in July 1881; water from Lake Vyrnwy first reached Liverpool in 1891; and the works were formally opened by the Duke of Connaught in July 1892.[35]

The ground conditions were ideal for the construction of the gravity dam, which is 1175 ft long and has a maximum height of 145 ft from the foundation to the crest of the spillway section. At this point the dam is about 127 ft wide across the base. This was the first high dam designed to act as a weir and so dispense with a separate spillway.

The Engineers, Deacon and Hawksley, adhered to the two fundamental principles of great weight and water-tightness, and much care was taken to achieve these objectives. The mass of the dam consists of large irregularly shaped stone blocks, up to 10 tons weight, set close together bedded on cement mortar. The spaces between them were then filled with mortar into which smaller broken stones were forced.

An important feature of the design was the provision of a drainage system in the foundations to prevent any build-up of water pressure on the under-side of the base, which build-up might lead to overturning.

The aqueduct is 68 miles long and there are now three 42 in. dia. pipes to deliver up to 50 million gallons per day to Prescot Service Reservoirs east of Liverpool. The route follows the Dee/Severn watershed to maintain high ground until the basins of the Mersey and Weaver are reached.

The first two pipelines were generally of cast iron but the use of rivetted steel pipes to facilitate maintenance in the 9 ft dia. cast iron

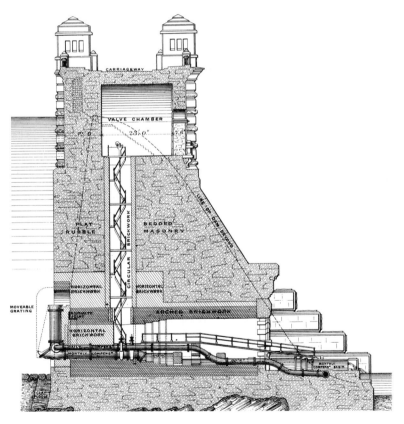

Vyrnwy Dam: cross-section

tunnel under the Mersey marked an early use of steel for trunk mains.

The first three tunnels, at Hirnant, Cynynion and Llanforda, are alike, with brick and concrete linings to protect against leakage. Hirnant was later duplicated at Aber to enable maintenance to be carried out.

At the Oswestry Reservoir, where the water is filtered, a 500 yd long earth dam impounds some 52 million gallons storage.

There are several balancing reservoirs and water tanks. The largest tank, at Malpas, has a capacity of $4\frac{1}{2}$ million gallons, while the Norton tank (HEW 1148, SJ 554 817), of 650 000 gallons capacity, is housed within a monumental sandstone tower some 110 ft high.[36]

The first section of the third pipe was laid in 1926–38 in steel. This saw the beginning of the more general use of bituminous coated steel pipes for trunk water mains in place of cast iron, which had been in use since 1810.[37]

After 1946, to increase capacity, a fourth pipeline was laid upstream of Oswestry with three booster stations downstream thereof.[38]

The pipe crossings under the Mersey and the Manchester Ship Canal were rearranged in 1978–81.

Horseshoe Falls Weir, Llantysilio

30.   LLANGOLLEN ANCIENT BRIDGE   HEW 165
SJ 215 422

This example of early road engineering is sometimes described as one of the Three Jewels of Wales.

The bridge, built on this site in 1282, was rebuilt and reconstructed in sandstone about 1500 AD to the present style with four arches, three of which are pointed and the fourth segmental. It has piers with cutwaters that extend up to the parapet walls forming recesses. In 1865 it was extended across the railway by an additional span of iron girders and dressed freestone.

The bridge was widened on the upstream side in 1873 to 20 ft between parapets and again in 1969 to 36 ft.

31.   SHROPSHIRE UNION CANAL,   HEW 1204
LLANGOLLEN BRANCH   SJ 196 433 to SJ 370 318

By the time the decision had been taken to change the originally projected north – south route of the Ellesmere Canal (HEW 1202, Chapter 8) to the more easterly route, progress had already been made up the Dee valley towards Llangollen, including the building of the Pontcysyllte Aqueduct (HEW 112) to take the canal across the river. The branch was completed in 1805 with an extension to Llantysilio under an Act of 1804 to obtain an adequate water supply, which is fed along the main line to the reservoir at Hurleston (SJ 626 553) on the Chester Canal (HEW 1202, Chapter 8), a fall of some 123 ft. The Chester Canal had drawn its supply from a feeder above Bunbury Top Lock, but this was scarcely sufficient.

It is interesting to speculate that if the change of route of the Ellesmere Canal had been decided upon earlier there might have been no need to build the magnificent Pontcysyllte Aqueduct, as the water supply could have been taken down the right bank of the Dee instead of the left.

32.   HORSESHOE FALLS WEIR,   HEW 1235
LLANTYSILIO   SJ 196 433
Illustrated on facing page

The Horseshoe Falls Weir was built by Thomas Telford to secure a water supply from the River Dee at Llantysilio to feed the

Ellesmere Canal. It is not clear whether the name derives from the shape of the weir, which is actually J-shaped, the horseshoe bend in the river, or even the pass through which runs the corresponding bend in Telford's Holyhead Road above Llangollen, from which there is an excellent view of the weir and its picturesque surroundings.

The weir is sited above rapids which are now an international venue for canoeing events. It is of masonry, 460 ft long with an upstream slope and a vertical downstream face 4 ft high. The crest is of 4 ft square stones with a flat bullnosed cast iron capping in 9 ft lengths, each length secured to the masonry by three lugs. A spare length of capping is stored on site.

The Horseshoe Weir is part of one of the earliest river regulation schemes in Britain, the need for which Telford recognized when in 1808 he obtained permission from the local landowner to raise the level of Llyn Tegid (Lake Bala) 2 ft 6 in. by means of a weir with sluices to ensure an adequate reservoir. Similar arrangements were made at Llyn Arennig Fach and Llyn Arennig Fawr.

Much of the water from the River Dee now goes to water treatment plants for public supplies as well as for canal impounding, and in more recent years the security of supply for Deeside and Merseyside has been augmented by the construction of reservoirs at Llyn Celyn, SH 936 356 (1965) and Llyn Brenig, SH 822 420 (1976).

---

33. PONTCYSYLLTE AQUEDUCT HEW 112
Illustrated on facing page SJ 271 420

Pontcysyllte aqueduct carries the Llangollen Branch of the Ellesmere (Shropshire Union) Canal over the River Dee two miles west of Ruabon and was commenced in 1795 and completed in 1805. It was scheduled an Ancient Monument in 1958.

The project attracted great admiration throughout the whole country and Robert Southey wrote of 'Telford who o'er the Vale of Cambrian Dee aloft in Air at giddy height upborne carried his navigable road', whilst Sir Walter Scott described it as the greatest work of art he had ever seen.

Previous canal aqueducts had consisted of a heavy puddled clay waterway constructed on squat brick arches, but Telford's scheme resulted in the novel use of cast iron, giving a trough 11 ft 9¾ in. wide carried on four arch ribs over each of the nineteen 44 ft 6 in. spans between the masonry piers.

The towpath, which has a protective iron parapet railing, overhangs the channel on the east side, thus giving space for the movement of water displaced by the passage of boats, while leaving a clear width of 7 ft 10 in.

The slender masonry piers, which are partly hollow, taper upwards to 13 ft × 7 ft 6 in. at the top. There is an interesting commemorative tablet fixed to the base of the pier adjoining the south bank of the river.

The embankment at the south end was one of the greatest earthworks undertaken at the time. The aqueduct is 1027 ft long and rises 121 ft above the river.

The aqueduct is important historically, first because Telford was working under the general supervision of William Jessop,[39] himself a pioneer in the use of cast iron, and secondly since it brought together Thomas Telford, Mathew Davidson his supervising engineer, William Hazledine the local iron master and the two master masons, John Simpson and John Wilson. Members of this team were subsequently engaged on many famous civil

Pontcysyllte Aqueduct (Ironbridge Gorge Museum Trust)

engineering works ranging from the Caledonian Canal to the Menai Bridge.

Many pictures of the Aqueduct show in the foreground the Pont Cysyllte (HEW 160, SJ 268 420), a three-span sandstone arched bridge built in 1696 across a bend in the river.

Visible from the Aqueduct is the Froncysyllte Bridge (HEW 1304, SJ 272 413) one of the 'Van Gogh' type lifting bridges which are a feature of the Ellesmere Canal (HEW 1204).

---

34.   CHIRK AQUEDUCT   HEW 111
SJ 286 371

This major work carries the Llangollen Branch of the Ellesmere Canal (HEW 1204) some 70 ft above the valley of the River Ceiriog. It was built between 1796 and 1801.

The structure appears from the exterior to consist of masonry piers and arches giving clear spans of 40 ft and is 710 ft long overall.

Telford, however, used cast iron plates to form the bed of the channel five feet deep, thus enabling him to reduce the depth of the masonry, and hence the weight, above the piers.

The plated bed was flanged at the edges and secured by nuts and bolts at each joint, and in addition was built into the masonry at each side. Side plates were added circa 1870.

The sides of the waterway were waterproofed by ashlar masonry and hard burnt bricks in Parker's cement, thus obviating the use of clay puddle used by the earlier canal engineers.

The adjacent 16 span railway viaduct (HEW 602, SJ 286 372) was built in 1848 by Thomas Brassey for the Shrewsbury and Chester Railway.

---

35.   CHIRK CANAL TUNNEL   HEW 162
SJ 285 376

Chirk Tunnel, at the north end of Chirk Aqueduct, built by Thomas Telford in 1801 with a length of 459 yd, is the longest of the three tunnels on the Llangollen Branch Canal. All three were unusual for the time of their construction in having a towpath taken through the bore, thus obviating the need for 'legging'. The portals make the tunnel appear larger than it is, as the bore is flared at the ends.

36. CEFN RAILWAY VIADUCT   HEW 568
SJ 285 412

This handsome viaduct, built by Thomas Brassey to the designs of Henry Robertson in 1846 – 48, carries the double line of the former Shrewsbury and Chester Railway (later the Great Western Railway, but which never became broad gauge) across the River Dee and adjacent fields about a mile downstream from the Pontcysyllte Aqueduct. The abutments and piers are of stone and the arches of brick with stone facings. There are 19 openings of 60 ft span and two of 15 ft. The total length is 510 yd and the greatest height 148 ft. It was claimed to be the longest viaduct in Britain at the time of its building.

CHAPTER 2

Mid-Wales

The area covered by this chapter lies eastward from the shores of Cardigan Bay to the English border and is bounded to the north by the River Dovey and to the south by the mountain ranges beyond which lie the valleys of South Wales. It is predominantly a land of mountains, forests and rivers. Both the Wye and the Severn have their birthplace on the slopes of Plynlimon and flow south-east and north-east, respectively, through the region.

The landscape and high rainfall of the area have attracted the attention of the water supply engineers, and man-made reservoirs

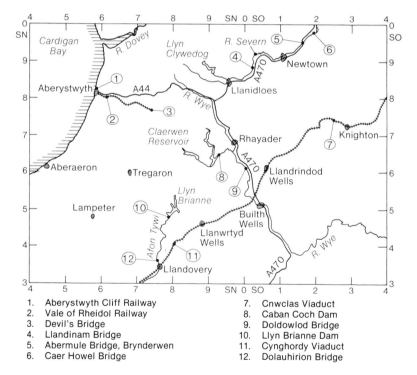

1.	Aberystwyth Cliff Railway	7.	Cnwclas Viaduct
2.	Vale of Rheidol Railway	8.	Caban Coch Dam
3.	Devil's Bridge	9.	Doldowlod Bridge
4.	Llandinam Bridge	10.	Llyn Brianne Dam
5.	Abermule Bridge, Brynderwen	11.	Cynghordy Viaduct
6.	Caer Howel Bridge	12.	Dolauhirion Bridge

abound. Of particular note is the Elan Valley group to the east of Rhayader, which supplies Birmingham.

Sheep and cattle farming are the major industries but there are both inland and seaside resorts, some served by the remaining rail network, whilst traces of ancient industries lie near to the occasional modern factory. Most of the works described in this chapter are associated with the provision of transport facilities in an area of rough topography and fast-flowing rivers.

1.   ABERYSTWYTH CLIFF RAILWAY   HEW 1130
Illustrated below   SN 583 826

The Cliff Railway at the northern end of Aberystwyth promenade was built for the Aberystwyth Improvement Company and opened on 1 August 1896. It is a funicular railway. The two passenger cars run on parallel 4 ft $8\frac{1}{2}$ in. gauge rail tracks and each is attached to a continuous wire rope cable which is operated by a stationary electric motor.

Aberystwyth Cliff Railway: car ascending from lower station

Devil's Bridge

The cars each have a capacity of about 30 persons. Their self weights are mutually balancing, one car being raised as the other is lowered. Until 1921 the railway was operated by a water balance system.

An early photograph indicates that substantial earthworks were involved in achieving an approximately uniform gradient for the track, which rises nearly 400 ft over its length of 798 ft.

| 2. | VALE OF RHEIDOL RAILWAY | HEW 1121 SN 585 816 to SN 739 770 |

The Vale of Rheidol railway, $11\frac{3}{4}$ miles long, is a 1 ft $11\frac{1}{2}$ in. gauge line, built in 1902 to transport lead and zinc ore from the area around Devil's Bridge to the harbour at Aberystwyth. It is now solely a passenger line and operates only during the summer months when it carries many thousands of tourists through the scenic beauties of the Rheidol Valley. Haulage is by steam locomotive and in fact the three units in use, named Prince of Wales, Owain Glyndwr and Llywelyn, are the only steam locomotives in service with British Rail. Ownership of the line passed to the Cambrian Railway in 1913, and to the Great Western Railway in 1922; it was nationalized in 1948.

3. DEVIL'S BRIDGE   HEW 1120
Illustrated on facing page   SN 742 771

Situated near Devil's Bridge station, the upper terminus of the Vale of Rheidol railway (HEW 1121), Devil's Bridge is one of the most popular tourist attractions in mid-Wales. Three bridges cross the Afon Mynach, one above the other. The first and lowest is a mediaeval pointed arch in stone of 15 ft span, traditionally thought to have been built by the monks of Strata Florida Abbey. The second bridge is considered to date from the mid-eighteenth century and is a flat segmental arch of 32 ft span.

Early in the 19th century the spandrel walls surmounting this arch were raised to allow the approaches to be less steep and at the beginning of the 20th century a steel bridge was built some 7 ft above the earlier structure. This bridge, since repaired and strengthened, continues to carry traffic to the present day.

Access into the deep gorge is though turnstiles via steps and pathways from which the bridges and nearby waterfalls can be seen. Motorists must use car parks in the vicinity.

4. LLANDINAM BRIDGE   HEW 850
         SO 025 886

This cast iron arch road bridge over the River Severn was built in 1846. The single arch spans 90 ft with a rise of 9 ft and is made up of three ribs, each consisting of five segments each with five X

Abermule Bridge

panels. The spandrels are also of open X pattern 4 in. cruciform members, generally similar to those of the Mythe Bridge at Tewkesbury (HEW 134, Chapter 5). The bridge is stiffened laterally by rectangular and circular cross-members connecting the arch ribs. It has a width of 10 ft 5 in. and carries a minor road. It was the first cast iron bridge to be built in the county of Montgomery and was cast by Hawarden Ironworks to the requirements of Thomas Penson, County Surveyor, who also designed the bridges at Abermule and Caer Howel.

5.   ABERMULE BRIDGE, BRYNDERWEN   HEW 342
Illustrated on page 43   SO 162 951

The elegant bridge over the River Severn at Abermule was cast by the Brymbo Company iron foundry in 1852. The single 110 ft span arch, with a rise of 12 ft 3 in., consists of five 2 ft 9 in. deep ribs, each with seven bolted segments. The abutments are of stone. An inscription cast into the two outer arch ribs records that it was the second iron bridge in the county of Montgomery. The width of 21 ft between parapets includes a 17 ft carriageway.

6.   CAER HOWEL BRIDGE   HEW 851
                                                           SO 197 982

A third cast iron bridge over the River Severn is also by the Brymbo Company, and rather later (1858). It has two spans of 72 ft 8 in., each with five ribs with five segments.

7.   CNWCLAS VIADUCT   HEW 594
                                                           SO 250 742

In the late 1850s, several railways were proposed to link the lines of South Wales with the West Midlands. One of these was built as the Central Wales line running from Craven Arms to Llandovery.

This is in picturesque country. The Cnwclas Viaduct carries the single line of the railway over a side valley of the River Teme as it starts its climb to the summit at Llangynllo Tunnel. The viaduct, designed by Henry Robertson and built by Thomas Brassey, is an elegant structure. There are thirteen arches of 30 ft span, the maximum height above the valley being 75 ft. Semi-circular stone arches surmount stone piers and the whole viaduct is richly decorated with castellated stone towers at each end and a battlemented parapet.

The section of line on which Cnwclas Viaduct stands was opened on 10 October 1863.

---

8.   CABAN COCH DAM   HEW 550
SN 92 64

In 1892 the Birmingham Corporation Water Act authorized the construction of reservoirs in the Elan and Claerwen valleys, southwest of Rhayader, and of an aqueduct (HEW 1194, Chapter 7) to convey water to the City. James Mansergh was the Engineer.[1] The initial works, in the Elan Valley, comprised three reservoirs, which were built by direct labour and completed in 1904.

Caban Coch is the first dam up the valley from Rhayader. It is 610 ft long, 122 ft high and 5 ft wide at the crest and is built of cyclopean mass concrete faced with block-in-course masonry. The downstream face has an inwardly curved batter, struck to a 340 ft radius, to 15 ft from the top, from which point to the crest the curvature is reversed. The area of the reservoir is 500 acres with an impounding capacity of 7815 million gallons. A novel feature of the scheme was the submerged dam built across the reservoir at

Doldowlod Bridge

Garreg Ddu, about $1\frac{1}{2}$ miles upstream from Caban Coch. This keeps the water at the required level to feed the aqueduct, while the part of the reservoir below it can be used to provide compensation water.

9. DOLDOWLOD BRIDGE   HEW 1245
Illustrated on page 45   SO 003 617

This privately owned suspension footbridge over the River Wye about five miles south of Rhayader lies within the Doldowlod Estate of the Gibson-Watt family. The suspension chains span 120 ft between 14 ft 6 in. high cast iron towers, each of which is built up from five separate castings bolted together, and each leg has an upper and lower section with a decorative cross-member. The links of the chains consist of $\frac{3}{4}$ in. dia. iron rods 7 ft 4 in. long, the number of rods in each link varying from five at the towers to two at mid-span. The links are connected by 1 in. dia. bolts. Inclined hangers attached to the chains at each link connection support the timber deck, an arrangement similar to that used by Dredge for his suspension bridges such as that over the Kennet and Avon Canal at Stowell Park (HEW 811, SU 146 614).

The bridge was built about 1880, the exact date being uncertain. The ironwork was cast at Llanidloes Railway Foundry.

Llyn Brianne Dam (B. I. G. Barr)

10.   LLYN BRIANNE DAM   HEW 552
Illustrated on facing page   SN 79 48

Llyn Brianne Dam and Reservoir lie in mountainous country to the west of Llanwrtyd Wells and north of Llandovery.

The dam, which was completed in 1972, stores and regulates the flow in the River Tywi in order that water may be abstracted at Nantgaredig some 40 miles downstream, to supply Swansea, Neath and Port Talbot.[2]

The crest is 900 ft long and 30 ft wide and the height of 300 ft makes it the highest dam in Britain. It is constructed of rock filling, with a boulder clay core. There is car parking provision for visitors who may drive some six miles along the length of the reservoir, and viewing points from which may be observed the spectacular jet by which water is discharged from the foot of the dam.

The work was designed by Binnie and Partners and constructed by George Wimpey & Co., Ltd.

11.   CYNGHORDY VIADUCT   HEW 272
SN 808 418

The Llandovery – Builth section of the Central Wales Railway line was opened in 1868. This section included a number of interesting engineering works, the viaduct across Afon Bran being the most spectacular. The viaduct has 18 spans of 36 ft and is curved in plan and 93 ft high. The single line railway is still in use.

The viaduct is about one mile from the village of Cynghordy, which lies on the road from Llandovery to Llanwrtyd Wells.

12.   DOLAUHIRION BRIDGE   HEW 167
SN 762 361

William Edwards, his two sons and his grandson, built a number of bridges in Wales in addition to the world-famous bridge at Pontypridd (HEW 27, Chapter 3). Some of these, like Pontypridd, were single-span bridges with openings in the haunches, and the best example is that at Dolauhirion, built in 1773. It is still in daily use although it has a weight restriction. The bridge spans the River Tywi approximately one mile from Llandovery and just off the road towards Rhandirmwyn and Llyn Brianne (HEW 552).

According to Jervoise the Dolauhirion bridge is the finest bridge over the upper part of the Tywi. He describes it as '. . . a single segmental arch with a span of 28 yards and a roadway 12 ft in width. The circular openings in the haunches of this bridge are a distinctive feature of Edwards' bridges'.

In the Dolauhirion bridge only one opening was provided at each abutment, and the feature is not, of course, unique to Edwards.

CHAPTER 3

South Wales

The industrial history of South Wales has been determined by its geography and geology, for beneath most of the area lies a bowl-shaped coalfield, outcropping in the north near to the borders of Powys and in the south along a line a little to the north of the M4 motorway.

In the 18th and 19th centuries ironworks grew up along the northern and, to a lesser extent, the southern edges of the coalfield to exploit the iron ore and limestone which accompanied the coal measures. Pack horse routes, canals and tramroads were established to take to the ports the products of the iron industry. From about 1840, as steam replaced sail, locomotives replaced horses, and improved techniques were developed for working the deeper seams, the output of coal, noted abroad as well as at home for its excellent steam-raising qualities, grew enormously. This led to the building of large docks at Newport, Cardiff, Barry, Swansea and elsewhere to cater for the massive export trade and to the growth of a railway system second only in extent to the network around London.

All this was not achieved without great effort, for the coalfield is crossed by a series of deep narrow valleys running approximately north and south, in which construction is fraught with difficulties. The canals in the valleys sloping steeply towards the sea had locks which frequently were deeper than was the practice in England, and the railways required numerous viaducts in timber, iron and masonry across the valleys, with tunnels through the intervening mountains. The ports had to contend with some of the world's largest tidal ranges.

The advent of oil-firing and the internal combustion engine at the beginning of the 20th century began the decline of the coal trade. Much of the coal still being produced goes to steelworks and power plants which with petro-chemical complexes are concentrated along the seaboard, while the valleys strive to attract new light industries. The main emphasis on communications has changed from north-south to east-west and the ports have been redeveloped to handle other cargoes in place of their coal exports.

The coal mining industry has preserved few remains of its past. Today civil engineering techniques are in use in the great opencast coal sites which fringe the coalfield and in the clearance of the many hundreds of abandoned spoil tips.

Lack of space in the narrow valleys has often meant that new works have had to be built virtually on top of the old, and miles of (often disused) canal and railway now lie between modern road-works, which in many cases have involved works as outstanding as those of the past, a great many of which remain as a testimony to the extent of the development which took place so rapidly in the 19th century.

1. THE LONDON TO   HEW 1198
FISHGUARD ROAD   SO 832 185 to SN 414 200

Communications between Great Britain and Ireland, with the associated sea links, became especially important in the early part of the 19th century. Telford's work on improving the Holyhead road is described elsewhere in this book, as are other roads along the North Wales Coast and the railway alternatives, the erstwhile Grand Junction Railway, the Chester and Holyhead Railway, etc.

What is now the A48 road across South Wales also played some part in the Irish traffic in connection with the Milford – Waterford

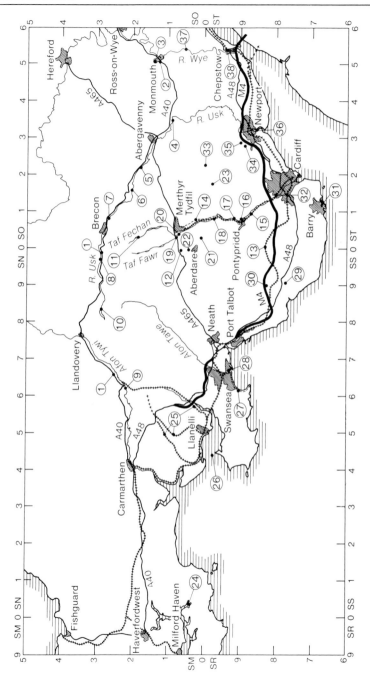

service until this was closed down in 1848. The A48 continues to link the coastal towns from Chepstow to Carmarthen, though its successor, the M4 motorway, now takes the bulk of the through traffic.

The A40 trunk road takes a more northerly route from Gloucester via Abergavenny and Brecon. As so often, the Romans had a hand in it. They linked their important centres at Gloucester and Caerleon, via Chepstow and, to avoid the lower Wye, via Monmouth. Another road ran westward to join Caerleon to Carmarthen.

From Gloucester, where until recently the road used Telford's bridge at Over (HEW 148, Chapter 5), the A40 goes westward to Ross-on-Wye and there turns south-west towards Raglan, entering Wales (Gwent) some two miles north of Monmouth where it passes near to the Monnow Bridge (HEW 146). This stretch of dual carriageway is part of the main highway from Birmingham to South Wales, via the M5 and M50 motorways and the A449 trunk road, which turns off at Raglan to join the M4 motorway east of Newport.

From Raglan the A40 goes westward and northward, and after Llanfihangel it follows the north side of the River Usk all the way through Abergavenny and Brecon, at a height of 100 to 200 ft above sea level.

At Abergavenny the Heads of the Valleys Road, the reconstruction of which began in 1960 before the period of large-scale motorway building, turns off to Merthyr Tydfil and on down the Vale of Neath to join the M4 at Neath.

About nine miles east of Brecon, the road swings briefly away from the river to climb steeply over a spur of the Black Mountains to reach a height of 660 ft at Bwlch. Shortly before Brecon the Brynich Aqueduct (HEW 337) may be seen just to the south of the road.

From Brecon to Senny Bridge the road is on the south side of the Usk but then turns away north-west through Trecastle, leaves the river and climbs to over 700 ft before descending the valley of the Afon Gwydderig to the 200 ft level at Llandovery, with Dolauhirion Bridge (HEW 167, Chapter 2) not far to the north. The sharp curves and steep gradients on this stretch no doubt explain the roadside monument at SN 802 349 to passengers killed in a stagecoach accident in 1835.

From Llandovery to Carmarthen the road follows the valley of the Afon Tywi, passing to the north of the bridge at Llandeilo (HEW 681).

The A40 remains the main road from Carmarthen to the west of South Wales, but elsewhere its importance has declined now that the Severn Suspension Bridge (HEW 201, Chapter 4) gives a direct service to London and South-West England.

2.   MONNOW BRIDGE, MONMOUTH   HEW 146
Illustrated below   SO 505 125

The Monnow Bridge at Monmouth was built in 1272 and is the only bridge in Britain which still carries a fortified tower.

The bridge, totalling 114 ft in length, has three semicircular masonry arches, each having three wide ribs. It has been widened by some $3\frac{1}{2}$ ft on the upstream side and 5 ft on the downstream side to give a width between parapets of 24 ft. The widening has been carried out with great skill so that it blends well with the original structure.

The fortification is arched over the roadway above the eastern pier of the bridge. Surmounting the archway there is a room 36 ft long by 10 ft wide covered by a pitched roof.

Monnow Bridge

The bridge at Warkworth, Northumberland (HEW 696, NU 249 062),[1] has a fortified tower at its southern end but not on the bridge itself.

| 3. MONMOUTH FORGE | HEW 1018 |
| GENERATING STATION | SO 503 137 |

A short way along the A466 north of Monmouth, an unclassified road to the left leads towards Osbaston. About 300 yd along this road another turning to the left leads to a small industrial complex. Here is the site of one of the earliest of the public electricity supplies in Britain and one of the earliest stations in Great Britain to generate at high voltage at a time when most municipal generation was at 600 V for direct transmission to tramway networks.[2,3]

First commissioned in 1899, it had three alternators, each 7 kW, 3000 V 60 cycles single phase a.c., driven by water turbines under the brick built station, which supplied electricity for lighting Monmouth.

The station was on the site of the Monmouth Forge and Tinplate works, where there had been an ironworks from as early as 1628. Three hundred yards upstream the forge weir in the River Monnow was raised to provide an enlarged reservoir from which water was led to the station along a canal.

Each alternator had a standby steam engine in case of drought and a further steam engine (replaced by oil in 1923) drove a separate 21 kW generator. The system suffered initial teething troubles arising from lack of experience in the design of the several transformers scattered about the town to reduce the voltage to the consumers' voltage of 100/110 V. Originally established by the municipality, the station was sold to a private firm in 1930. After nationalization of the industry in 1948 power continued to be fed to the national grid until 1953. The reservoir and weir (HEW 1269, SO 502 138) are still in use as a river regulator. Of the rest of the works only the station building remains, housing a small factory.

| 4. PANT Y GOYTRE BRIDGE | HEW 704 |
| | SO 348 089 |

According to Jervoise, Pant y Goytre bridge was built in about 1821 and locally it is considered that the engineer was probably

John Upton, to whom is also attributed the nearby Llanellen bridge at SO 306 111. It carries the A471 road over the River Usk just south of the A40.

The bridge has a centre span of 58 ft and side spans of 39 ft. The arches are elliptical in form. The spandrel faces are pierced by cylindrical voids close to the springings, and between the springings there is a larger (9 ft dia.) void over each pier. Similar 9 ft dia. voids through each abutment also serve for flood relief. As in Halfpenny Bridge, Lechlade (HEW 628, SU 214 993) the masonry in the spandrel faces is laid with beds radial to the arch.

This is a most attractive bridge, graceful in shape and in beautiful surroundings.

---

5. CRICKHOWELL BRIDGE   HEW 1223
                                                                SO 214 181

Just south of the A40 between Crickhowell and Bwlch two rather splendid masonry 16th/17th century bridges cross the River Usk.

That at Crickhowell, which is a scheduled Ancient Monument, has 13 flat segmental arches ranging in span from 16 ft 10 in. to 39 ft 6 in. It was widened on the downstream side early in the 19th century but still can only carry single line traffic. Its size and its position on the A4077, which turns south-west off the A40 through the village, suggest that the main coaching road once crossed it to take a route close to that adopted for the canal between Brecon and Abergavenny.

---

6. LLANGYNIDR BRIDGE   HEW 1247
                                                                SO 151 202

Five miles west of Crickhowell the Llangynidr bridge carries the B4560 road, about ½ mile south of its junction with the A40. The six segmental arches vary in span from 22 ft to 30 ft 6 in. The arch rings have two courses of voussoirs, the outer course being shallower than the inner but overhanging it somewhat. The roadway is only 8 ft wide.

The bridge stands in a well-frequented beauty spot and, though shorter than the Crickhowell Bridge, it is of the two the more impressive, with its massive piers and its arches standing higher above the river. It is a Grade II listed building.

As at Crickhowell the bridge has triangular cutwaters to the piers, carried upwards to the tops of the parapets to provide pedestrian refuges.

---

7. BRYNICH AQUEDUCT   HEW 337
                                              SO 079 273

Brynich Aqueduct is a four-span masonry structure carrying the Brecknock and Abergavenny Canal over the River Usk, just south of the Abergavenny – Brecon road, part of the A40, about two miles east of Brecon. It was built in 1799 – 1800, the Engineer being Thomas Dadford, Jr.

It is approximately 210 ft in length. The canal width is 12 ft within an overall width of 31 ft and the water level of the canal is about 35 ft above river bed level. The spans are approximately 36 ft and the pier widths 7 ft 6 in.

---

8. ABERCAMLAIS SUSPENSION   HEW 1265
        BRIDGE   SN 965 290

In the grounds of the Abercamlais estate adjoining the A40 trunk road about six miles west of Brecon, a slender suspension bridge of 80 ft span crosses the River Usk. Built circa 1830 it is attributed to Crawshay Bailey, the famous ironmaster of Nantyglo in the eastern Ebbw Vale.

Two $1\frac{1}{8}$ in. dia. wrought iron rods on each side, spaced 2 ft 8 in. apart vertically form the suspension cables, their screwed ends passing through cast iron end posts, 6 ft high and spaced 2 ft 2 in. at the bottom and 3 ft 4 in. at the top, to which they are tensioned by nuts. Every 10 ft, iron rods hooked over the top cables and looped round the lower, carry transverse 4 in. by $\frac{3}{8}$ in. flats as deck-bearers. Between these short hangers attached to the lower cables carry $1\frac{1}{2}$ in. by $\frac{1}{2}$ in. bearers. The footway is of four 4 in. by $\frac{3}{8}$ in. iron flats spaced to give a width of 1 ft 6 in. and rivetted to the wide deck bearers but resting freely on the others. The bridge has been kept in excellent condition by the owners.

At Rhosferrig (SO 033 515) near Builth Wells there was a similar but somewhat larger bridge (HEW 1184) with three suspension cables on each side, also attributed to Crawshay Bailey. Originally built over the Usk near Crickhowell circa 1830 it was transferred to its later site over the Irfon in 1836. Unfortunately through age and flood damage it deteriorated beyond repair and

Powys County Council has replaced it. Parts of the old bridge are preserved at the Welsh Industrial and Maritime Museum at Cardiff.

Just upstream from the Abercamlais bridge is a four-span arched masonry bridge which carries an estate road across the Usk. It was built contemporaneously with the house as a packhorse bridge circa 1600 and widened a century later to carry wheeled traffic.

9.   LLANDEILO BRIDGE   HEW 681
                                                      SN 627 220

Llandeilo Bridge has a clear span of 145 ft and an overall width of 33 ft. It was built in 1848 and carries the trunk road A483 on a falling grade from the town over the Afon Tywi and its neighbouring water meadows.

The bridge is of masonry, finished to a standard approaching that of ashlar work. Perhaps in style it is too formal for such a rural setting but it has been described as probably the finest one-arch bridge in Wales. There are interesting references in *The Diary of Thomas Jenkins of Llandeilo* to its construction, on which the diarist worked.[4]

10.   USK DAM   HEW 1240
                                                       SN 83 28

Not far from the source of the River Usk is an earth dam some 100 ft high and 1600 ft long on the crest, built in 1955 to create a water supply reservoir. It was designed by Binnie Deacon and Gourlay and built by Richard Costain Ltd. It is of historical interest because it was the first earth dam in Britain to be provided with horizontal drainage blankets in the embankment, a practice suggested by Professor A. W. Skempton of Imperial College, London, and which has since become standard when clay fill is used.

11.   UPPER NEUADD DAM   HEW 1241
                                                       SO 03 18

There are six reservoirs on the Taf Fawr and Taf Fechan which flow from the Brecon Beacons to Merthyr Tydfil to form the River

Taff. One of these is the Upper Neuadd, built in 1902 to the design of G. F. Deacon. The dam is of the gravity type, 77 ft 5 in. high and 1385 ft long on the crest, and is unusual in that the entire dam is of a thin masonry section buttressed with earth fill over most of its length.

12.	NANT HIR RESERVOIR	HEW 1305
		SN 98 06

Another tributary of the Taff, the Afon Cynon, has several interesting reservoirs on it above Aberdare. One of these, built in 1875 on a side stream at Nant Hir, has an earth dam some 63 ft high and 328 ft long on the crest, with a puddle clay core, a grassed downstream slope and stone pitched upstream slope.

It was the work of J. F. La Trobe Bateman, one of the most active water engineers of his day.[5] Other examples of his work are mentioned in Chapters 5 and 8. He was the son-in-law of William Fairbairn whose work on tubular bridges in association with Robert Stephenson is referred to in Chapter 1.

Across the Nant Hir reservoir is a notable reinforced concrete open-spandrel arch bridge carrying the A465 Heads of the Valley Road.

13.	SOUTH WALES RAILWAY	HEW 1199
		ST 536 937 to SN 412 196

The Great Western Railway broad gauge (7 ft $0\frac{1}{4}$ in.) lines were extended into South Wales in 1851 via Gloucester and Chepstow (HEW 1179). This rail route almost coincides with the road, now designated A48, which diverges, as it did in Roman times, from what is now the A40 at Over, and rejoins it at Carmarthen, linking Newport, Cardiff, Neath and Swansea on the way.

Apart from the bridge over the Wye at Chepstow, notable structures on the South Wales Railway were the crossing of the Usk at Newport, subsequently rebuilt and widened; Landore Viaduct (HEW 327, SS 663 959)[6] whose many timber spans were replaced by iron in 1888 and by steel in 1979; Llwchwr Viaduct (HEW 1131, SS 561 980), another timber viaduct whose deck is now in steel; and a drawbridge opening span at Carmarthen built by Brunel in 1854 and rebuilt 1910 as a Scherzer rolling lift bascule bridge (HEW 1216, SN 405 192).[7] Models of the main span of the original Landore Viaduct are in the Royal Institution Museum in Swansea and the National Railway Museum at York.

The curves at Neath are not conducive to high speed running and the few stiff gradients thereabouts made heavy demands on motive power. This situation was eased in 1913 by the construction of a by-pass line which crossed the River Neath by an interesting swing bridge (HEW 1232, SS 730 964), originally built by the Rhondda and Swansea Bay Railway. Five fixed spans of plate girder construction, varying in length from 40 ft to 52 ft 6 in. have been renewed in recent years. The movable span is a Pratt truss with a curved upper boom, swinging about its centre and resting on a cast iron roller race, the span being 167 ft 6 in. Originally operated hydraulically the bridge is now fixed. It is the only opening bridge of this type in Britain built both on the skew and on a curve.

Finch of Chepstow supplied the steelwork; Sir William Armstrong of Newcastle the hydraulic machinery.

At Llansamlet, between Neath and Swansea, four 70 ft span stone arches (HEW 802, SS 702 925) which may be compared with the Chorley Flying Arches (HEW 751, SD 575 195)[8] spring from the sides of a cutting. Brunel seems to have built them to permit steeper slopes and so save on excavation.

Bridgend Station was overhauled in 1980 and is now a pleasant mixture of the original Brunel buildings (listed) and modern railway architecture, which received awards from HRH the Prince of Wales and from the Development Corporation of Wales.

The South Wales Railway was originally intended to lead to a cross-channel port at Fishguard, but economic conditions in Ireland at the time defeated this object. Fishguard was not in fact developed seriously until 1904.

The use of the broad gauge after it had been officially declared non-standard in the United Kingdom was unfortunate, since the huge output from the South Wales collieries demanded mineral wagons which would be capable, not only of working over any part of the English rail network irrespective of track gauge, but up the Welsh valleys themselves in an economic manner. Even Brunel had thought in 1840 that something smaller than the broad gauge was desirable there when he built the Taff Vale Railway (HEW 1219). He played this down when giving evidence to the Gauge Commission in 1845, but had to admit that he did recommend 4 ft $8\frac{1}{2}$ in. for the Genoa-Turin Railway because of interchange traffic requirements with adjoining 'narrow gauge' lines.

By 1866, so great was the clamour in South Wales to convert the main land exit route eastward into England to standard gauge, that this was achieved by 1872.

The long detour via Gloucester was still a disadvantage for London, Bristol and West of England traffic. Brunel's original proposals had indeed envisaged a crossing of the Severn at Hock Cliff (SO 730 090) but this was vetoed by the Admiralty. Later, he made a more direct link near Chepstow (HEW 1033), but it was not until 1886 that the GWR solved the problem by the construction of the

Penydarren Tramroad: replica of Trevithick's locomotive (courtesy Welsh Industrial and Maritime Museum)

Severn Tunnel (HEW 232) and, for London traffic, by a direct link from it to Swindon in 1903 (HEW 1073). (See Chapter 4.)

The M4 motorway was extended into Wales following completion of the Severn Suspension Bridge (HEW 201) in 1966 and only one section, west of Port Talbot, remains to be constructed. The three routes, M4, A48 and the railway, are seldom far apart from each other between Chepstow and a point some 14 miles short of Carmarthen, where the two roads merge.

14. PENYDARREN TRAMROAD   HEW 705
Illustrated on facing page   SO 056 070 to ST 085 950

The Penydarren Tramroad, engineered by George Overton and opened in 1802, has a unique place in railway history. On it, on 21 February 1804, what is generally regarded as the first journey by a steam locomotive was made by a machine built by Richard Trevithick. The L-section cast iron rails were 3 ft long, weighing 56 lb each, laid on stone blocks to a gauge of 4 ft 4 in. over the outsides of the flanges. The rails proved too fragile for the weight of the locomotive, so that although the experiment showed the possibilities of this new form of traction it also demonstrated that these could not be fully realized until a much better track became available – which did not happen until 1825, on the Stockton and Darlington Railway (HEW 85).[9]

The Welsh Industrial and Maritime Museum at Cardiff has a working replica of the locomotive. The rails on which it runs are replicas of the Tramroad rails, but laid on concrete blocks instead of stone. A modification to the seating of the rails on the blocks has, by reducing the effective span, overcome the breakage problem.

The Glamorganshire Canal was completed between Merthyr Tydfil and the sea at Cardiff in 1798. In consequence of disagreements between Richard Crawshay, whose Cyfarthfa ironworks were best served by the canal, and the other ironmasters of Merthyr, a tramroad was constructed from Samuel Homfray's Penydarren works to a point on the canal at Abercynon (the Navigation Hotel nearby was once the offices of the Canal Company). This route effectively bypassed part of the canal route, particularly ten sets of locks north of the junction at Abercynon.

From its southern terminus at Abercynon (ST 085 950) the route of the Tramroad runs along the eastern bank of the River Taff. Near Quaker's Yard the river was crossed by timber bridges at ST 094 963 and ST 090 966 (HEW 799). About 1815, these were

replaced by segmental, nearly semicircular, masonry arches of
60 ft span and 9 ft wide, which have parallel rings about 18 in.
deep composed of 2 to 3 in. flat stone voussoirs similar to Pont y
Gwaith (HEW 800), and indeed, Pontypridd (HEW 27). The old
alignment can still be traced.

Further on, the Tramroad was eventually crossed by the Taff
Vale Railway on a viaduct (HEW 801, ST 089 965). The section
between Quaker's Yard and Pont y Gwaith follows the deep valley
of the Taff through pleasantly wooded country.

North of Merthyr Vale the route crosses to the east side of the
A470 trunk road and continues to Merthyr where its course is
perpetuated in the names of two streets, Tramroadside South and
Tramroadside North.

At SO 056 070, near the point of termination, there is a
memorial to Richard Trevithick.

---

15.   TAFF VALE RAILWAY   HEW 1219
SO 052 057 to ST 190 750

At Pontypridd the River Taff is joined by the River Rhondda
before flowing on to Cardiff. Between the two rivers lies the valley
of the Cynon, a tributary of the Taff.

At the turn of the 18th–19th centuries the discovery of iron ore
along the heads of the valleys and coal in the valleys led to the
establishment of ironworks, as at Cyfarthfa and Dowlais near
Merthyr Tydfil, and to a rapid expansion in coal mining. These
industries required means of transport for the raw materials and
for the finished products. The Glamorganshire Canal from Merthyr to Cardiff was built between 1790 and 1798 to give communication with the sea, but in general the South Wales coun-
tryside did not really lend itself to the development of canals, and
so there was a great proliferation of horse tramroads such as
Penydarren (HEW 705).

By 1830 canals and tramroads were proving inadequate, so in
1835 the Merthyr ironmasters engaged I. K. Brunel to construct
the first major commercial railway in South Wales. An Act of June
1836 authorized a single 4 ft $8\frac{1}{2}$ in. gauge track, $24\frac{1}{4}$ miles long,
with six passing places, from Merthyr, at the head of the Taff Vale,
to Cardiff, where dockland development was beginning.

Major works included a rope-worked incline at 1 in 19 north of
Navigation (Abercynon); a masonry viaduct at Pontypridd (HEW
1220, ST 071 900) with a skew span of 110 ft across the Rhondda;

Pontypridd Bridge

and the six span masonry viaduct across the Taff and the Penydarren Tramroad at Quaker's Yard (HEW 801). This is unusual in having octagonal piers with deeply chamfered arches. There were three other crossings of the Taff, at Whitchurch and to the north and south of Taffs Well. The southernmost sixteen miles were opened in October 1840 and the line was opened throughout on 28 April 1841.

In the course of time, coal became the dominant traffic. The Taff Vale Railway expanded into the Cynon valley from 1845; by 1856 it had reached Treherbert near the head of the Rhondda Fach and in the following year the line to Merthyr had been doubled. Before the end of the century the company had penetrated to Llantrisant, Cowbridge and Aberthaw and beyond Cardiff to Penarth, where it eventually owned the docks, and to Barry. The railway now had adequate outlets to the sea, but, in spite of some not entirely satisfactory transfer arrangements built at Cardiff, was not able to make use of the broad gauge South Wales Railway (HEW 1191) to move its traffic eastward across the Severn into England until the conversion of that railway to standard gauge in 1872.

The Newport Abergavenny and Hereford Railway Co. exploited this situation to gain a share of the lucrative business by pushing their Taff Vale extension westwards to join the TVR at Quaker's Yard. It was over these routes that there was a massive movement of coal northwards in the First World War to fuel the Navy at Scapa Flow.

The Taff Vale, being first in the field, was able to choose the easiest routes in difficult country. Later railways competing for the lucrative coal traffic had to contend with a harsh topography, sometimes having to move from valley to valley, and at the expense of long viaducts and tunnels.[10] It is little wonder, then, that the surviving lines of British Rail in the area largely use the pioneering routes of the Taff Vale Railway.

16. PONTYPRIDD BRIDGE   HEW 27
Illustrated on page 63   ST 074 904

There are a number of interesting engineering works in the vicinity of the Taff and Cynon valleys north of Pontypridd.

No study of bridges – and especially masonry arch bridges – would be complete without a reference to William Edwards's famous arch across the River Taff at Pontypridd.[11,12] His first attempt, in 1750, a multi-span bridge, was washed away in a flood after only two years. Edwards then decided to span from bank to bank, but again his arch was washed away in a flood even before

Berw Road Bridge, Pontypridd

the centre was struck. His next attempt was also doomed to failure. The span of 140 ft involved a rise of 35 ft and this meant a great weight of filling over the haunches compared with the crown where there was only the arch ring and the parapets. During the construction of the spandrel walls the excessive weight near the abutments forced the crown upwards and the bridge again collapsed. Fortunately this was not a sudden failure and Edwards had time to observe the mode of collapse.

Jervoise has described the third and successful single span as follows: 'Edwards then rebuilt the bridge to the same design except that he placed at each end three cylindrical holes graduated in size, the largest being 9 ft in diameter, to relieve the arch from the pressure of its haunches'. The spandrel infilling was of charcoal, for further lightness.

This scheme proved successful, and the bridge, which was completed in 1755, still stands.[13]

The bridge soffit is an almost perfect arc of a circle 89 ft in radius and the arch ring has a depth of construction of only $2\frac{1}{2}$ ft. The relatively large rise at the crown resulted in steep slopes at either end of the bridge and this caused serious problems for heavy carts – both during the ascent and descent. A modern bridge has been built alongside and Edwards's masterpiece is preserved and used for pedestrians only. It is 11 ft wide between parapets.

---

17.   BERW ROAD BRIDGE, PONTYPRIDD   HEW 620
Illustrated on facing page   ST 077 911

The Berw Road Bridge, which crosses the River Taff about half a mile upstream of William Edwards's masonry arch bridge, is one of several early reinforced concrete bridges in South Wales built on the Hennebique system.

The bridge has a central clear span of 116 ft and side spans of 25 ft. The width between parapets is 26 ft.

The main span has three parabolic arched ribs, at 12 ft centres, cross-braced at intervals. The longitudinal beams supporting the deck are supported by columns off the ribs over the outer thirds of the span and the arch itself serves as direct support to the deck over the middle third.

The side spans have their outer main beams arched to match the centre span.

The bridge was built in 1907 to the design of L. G. Mouchel and Partners. The deck was reconstructed in recent years.

18.   PONT Y GWAITH   HEW 800
ST 080 975

Pont y Gwaith is a masonry bridge of 55 ft span, 15 ft 9 in. rise, over the Taff about a mile north of Quakers Yard. It has several features in common with Pontypridd (HEW 27), including the use of thin stones to form the arch ring; the severe road gradient; and the narrowing in plan from abutments to mid-span; but has no opening in the spandrels. Both span and width of bridge are much less than those of Pontypridd. There is a noticeable tendency of the arch to be pointed at the crown like the first Ouse Bridge at York. At the present time the bridge is being affected by mining subsidence.

19.   CEFN COED Y CYMMER VIADUCT   HEW 171
SO 030 076

During the second half of the 19th century the network of railways expanded rapidly in South Wales to handle the vast tonnages of coal being produced. The difficult terrain forced the construction of numerous viaducts, some with masonry arches, some with metal spans. Often of great height and curved in plan, they produced some excellent and sometimes unique examples of bridge engineering. Many have since been demolished and others are disused, such as Cefn Coed, a masonry viaduct which carried a part of the Brecon and Merthyr and London and North Western Joint Railway over the Taf Fawr. It is 770 ft long and 115 ft high with 15 semicircular arches of 39 ft 9 in. and is on a curve. It was designed by Alexander Sutherland in consultation with Henry Conybeare and built by Savin & Ward in 1866.

The line was opened on 1 August 1867 and closed in 1962.

20.   PONT Y CAFNAU   HEW 656
Illustrated on facing page   SO 038 071

This unique cast iron 'bridge of troughs' still spans the River Taff where it was built in 1793 to carry a tramroad and water supply into the Cyfarthfa ironworks in Merthyr Tydfil.[14] The designer was the chief works engineer, Watkin George. In 1795 a second bridge, which no longer exists, was cast from the same patterns to carry an extension of the tramroad from the works to the Glamorganshire Canal.

The bridge, now used by pedestrians, spans 47 ft. Two substantial A-frames, one on each side of the deck, have their feet embedded in the river walls, with the apex at mid-span. The frames are held together by mortice-and-tenon and dovetail joints (George was a former carpenter) and incorporate sockets which carry transverse members at mid-span and at the quarter points. These in turn support the deck structure, which is a closed rectangular box about 2 ft deep and 6 ft 2 in. wide.

Pont y Cafnau undoubtedly had its influence on other, better known, aqueducts. In 1794 the Shropshire ironmaster, William Reynolds, sketched the bridge and in the following year Telford reported that the design and method of construction of Longdon-upon-Tern Aqueduct (HEW 280, Chapter 7), itself a prototype for Pontcysyllte (HEW 112, Chapter 1), had been referred to Reynolds and himself.

| 21. | **RAILWAY BRIDGE AT CWMBACH, ABERDARE** | **HEW 1055** SO 024 011 |

This steel truss bridge, which carries the only remaining railway line in the Aberdare valley over the Afon Cynon, played an

Pont y Cafnau

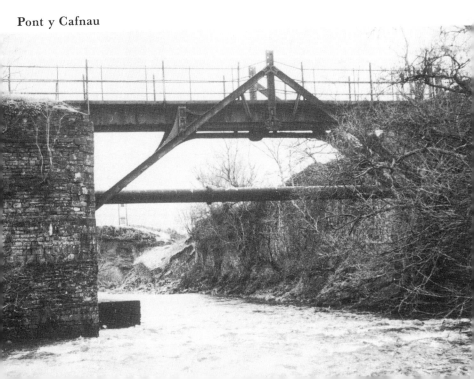

interesting role in the development of modern design practices and construction techniques.

In the 1960s, the A40 trunk road east of Oxford was being widened from single to dual carriageway, and at Wheatley a bridge was erected in 1961 to carry the single-track Oxford to Thame branch over the new carriageway. It was required to match in general appearance the adjacent rivetted truss bridge over the existing road.

P. S. A. Berridge was responsible for the design and chose to develop a structure which was largely prefabricated by welding, but with the necessary degree of site assembly carried out by Tor-shear bolts, which acted by applying a known pressure to the mating surfaces of the joints.

The experience gained was used in the construction of the much larger trusses, also designed by Berridge, which, in 1962, replaced I. K. Brunel's tubular suspension bridge at Chepstow.

Both bridges were fabricated at the Fairfield works at Chepstow, successors to Messrs. Finch, who did so much work for Brunel.

In 1973, on closure of the Thame branch, the bridge at Wheatley was dismantled and re-assembled near Aberdare. This time the more recently developed Huck fasteners were used for joint assembly.

Carew Tide Mill (J. P. Davies, Haverfordwest)

22. ROBERTSTOWN CAST IRON TRAM BRIDGE, ABERDARE

HEW 371
SN 997 037

This interesting little bridge is certainly one of the oldest surviving 'railway' bridges in the world. In 1811 the Aberdare Canal Company completed a tramway between Hirwaun and the canal head at Cwmbach. The bridge carried the tramway across the River Cynon between Trecynon and Robertstown.

Four arched and trussed cast iron beams spring from continuous cast iron brackets built into the abutments. The width of each truss is only 3 in. and the depth varies from 1 ft at the centre to over 5 ft at the ends. Seventeen cast iron plates 9 ft 11 in. wide make up the total length of deck of 36 ft 8 in. The stone abutments were built with obvious skill.

23. HENGOED VIADUCT

HEW 804
ST 155 949

The masonry Hengoed, or Maes-y-Cwmmer viaduct, like the iron viaduct at Crumlin (HEW 72) was one of the major structures on the Taff Vale extension of the Newport, Abergavenny and Hereford Railway, and was built in 1857. It has 16 spans of 40 ft semicircular arches and a maximum height of 130 ft. It is built in rough stone, on a curve, and crosses the A469 (spans three and four), River Ryhmney (span seven), and a minor road (span eleven).

24. CAREW TIDE MILL
Illustrated on facing page

HEW 803
SR 041 038

Carew Tide Mill, a large three-storey stone building, stands at the southern end of a dam across the estuary of the Carew River near the southern edge of Milford Haven.

The dam, which has stone facings and a central clay core, impounds an area of water of some 27 acres and this water is discharged through passages in the dam beneath the mill to operate undershot waterwheels, of wood and iron construction. Only one wheel is now in operation and it is considered to generate some 20 brake horse power.

Records of the mill can be traced back to Elizabethan times.

After a long period of disuse it was saved from imminent collapse in the early 1970s, largely through the efforts of Mr John Russell, FSVA, and with financial assistance from various organizations.

In the 1930s the sites of some 25 tide mills were recorded, mostly in southern Britain. Today few are left and fewer, among which may be mentioned that at Woodbridge in Suffolk (HEW 113, TM 275 497), are in working order.

---

25. THREE RAILWAYS IN DYFED

BURRY PORT AND GWENDRAETH	**HEW 1217**
VALLEY RAILWAY	SN 451 008 to SN 529 125
LLANELLI AND MYNYDD	**HEW 1388**
MAWR RAILWAY	SS 500 995 to SN 563 131
LLANELLI RAILWAY	**HEW 1389**
	SS 532 989 to SS 634 226

The area of Dyfed between the Rivers Llwchwr and Tywi and south of the A48 trunk road is rich in mineral resources, in particular high quality anthracite.

Three railways serve the area; all are based on the routes of early canals or horse-drawn tramroads which pre-existed the South Wales main line yet have survived to become branches of the major route.

Until very recent modifications which have eliminated the section south of the Gwendraeth Fawr, the Burry Port and Gwendraeth Valley Railway ran from the South Wales main line at Burry Port in a westerly direction before crossing from south to north under the main line at SN 442 007 and, after a further three miles of running almost parallel, turned north and then north-east to climb the Gwendraeth Fawr valley as far as Cwm-mawr. It was built in 1869 on the line of a former canal, dating partly from 1815 and partly from 1839. For much of this canal, including the construction of the inclines at Pont Henri (SN 476 092), Capel Ifan (SN 491 104) and Cwm-mawr (SN 529 125), James Green was the Engineer.

The railway also made use of a former masonry aqueduct over the Gwendraeth Fawr (HEW 1234, SN 428 054). This has six flat arches of about 9 ft clear span with their soffits some 11 ft below track level and an overall width of 40 ft. It originally incorporated an iron trough, which may still be in position beneath the rail formation, to contain the canal. There is a second aqueduct at SN 447 073.

There are several aqueducts in Britain which now serve as road bridges, but the former Cart aqueduct at Paisley (HEW 813, NS 494 634) is the only other known to have become a railway bridge.

Sir William Shelford and John Robinson were the Engineers for the conversion. Many of the original canal bridges were not modified; one, an elliptical arch, still survives at SN 447 006. Until recently the line operated with a substandard headroom of about 12 ft and special rolling stock and locomotives were used.

The second surviving line, formerly the Llanelli and Mynydd Mawr Railway, was built along the line of the former Carmarthenshire Railway or Tramroad. The latter was the second public railway to be authorized by an Act of Parliament, of 3 June 1802, the first having been the Surrey Iron Railway of 1801 (HEW 1387, TQ 255 753 to TQ 318 656). J. Barnes and T. Morris were the Engineers. The line ran north from Llanelli up the valley of the Afon Lliedi and on to limestone quarries near Castell-y-graig (SN 603 158). It was not a great econ mic success and was closed in 1844.

In the course of time, because of the growing demand for anthracite, the concept of a railway was revived and, by an Act of 19 July 1875, the L&MMR was constructed over part of the route of the earlier line, as far as Cross Hands, and opened in 1883.

Today the line, running via Sandy Gate Junction (SN 498 004), which once led to the BP&GVR, serves the National Coal Board's modern colliery at Cynheidre (SN 494 075).

The third survivor is the former Llanelli Railway, which ran eastwards from the town, following the flat land adjacent to the Llwchwr estuary to reach Pontardulais (SN 587 040) in 1839. Originally built as a horse-drawn tramway, it was relaid with heavier materials as a standard gauge railway and extended to Llandeilo in 1857. A later branch ran from Pontardawe to Swansea, crossing the neck of the Gower peninsula.

The Llanelli-Llandeilo section remains and is now used by passenger trains of the Central Wales line, while a branch at Ammanford serves collieries in the Amman Valley and the new drift mine at Betws (SN 640 125).

---

26.   WHITEFORD POINT LIGHTHOUSE   HEW 1256
Illustrated on page 72   SS 444 973

This has been described as the only wave-swept cast iron lighthouse in the British Isles. A scheduled Ancient Monument, it stands half a mile offshore on the northwest tip of the Gower Peninsula and looks across the Burry Inlet to Burry Port and Llanelli, whose harbours have now largely fallen into disuse. It was built in 1865 under the Llanelli Harbour Act, 1864, to replace an earlier light, carried on timber piles, which had lasted only a few years. The light was extinguished in 1921, but in 1982, with the aid of a grant from the Llanelli Harbour Trust, the Burry Port Yacht Club installed a solar-powered light in the lantern.

The lighthouse stands just above low water mark and is 61 ft high. Eight courses of cast iron plates, bolted together through

Whiteford Point Lighthouse (N. J. Francis, Llanelli)

internal and external flanges, form the circular tower which has the curved taper typical of many lighthouses. The lantern is surrounded by an ornate wrought iron balcony. At some time after erection the tower was strengthened by wrought iron bands.

The area to the south of the lighthouse, which is some two miles from the nearest road, is in the care of the National Trust and leased to the Nature Conservancy Council. The lighthouse is accessible on foot at low tide but because of unexploded shells and quicksands in the vicinity and the sensitive nature of the reserve, intending visitors should first get in touch with the Warden at the Council's Oxwich Reserve Centre, Oxwich, Gower.

| 27. SWANSEA AND | HEW 706 |
| MUMBLES RAILWAY | SS 645 921 to SS 629 874 |

The Oystermouth Railway, as it was originally known, was authorized by Parliament in June 1804 and was the world's first fare-paying passenger 'railway'.[15,16] The first load of passengers was carried on 25 March 1807, using horse traction. The Act mentions haulage by 'men, horses, or otherwise', which last word might just possibly have had Trevithick's Penydarren locomotive in mind.

Steam traction was introduced in 1877. In 1879 the line was renamed the Swansea and Mumbles Railway and in 1898 was extended to Mumbles Pier. It was electrified in 1928, being operated by double-decker tramcar type vehicles, the largest in use in Britain, and continued in service until 5 January 1960.

The track ran along the seaward side of the A4067 which links the Gower Peninsula to Swansea. Although the tracks have now been removed the course of the six mile route can still be followed, and the former Blackpill station, built in 1927, which also housed a substation, remains in good condition.

In March 1981 a stained glass window was unveiled in Oystermouth Parish Church to mark the 175th anniversary of the opening of the railway.

| 28. WEAVER'S MILL, SWANSEA | HEW 354 |
| | SS 661 931 |

Although Weaver's Mill was demolished in 1984 to make way for a supermarket, it will always be of interest since it was probably the first example of a multi-storey reinforced concrete frame industrial building to be built in Britain. The combined flour mill and granary, which was built in 1898, was constructed on the Hennebique System introduced from France in 1897, the designer being F. Hennebique, in association with L. G. Mouchel and H. C. Portsmouth, a Swansea architect.

The five upper floors of the seven-storey building, which was 80 ft by 40 ft and 112 ft high, cantilevered out from the main supports. Decorative mouldings were introduced to make the structure aesthetically more pleasing. The roof of the mill was used as a reservoir to hold 100 tons of water. Surprisingly, there was no continuity of reinforcement through the floors.

Three typical columns from the mill have been preserved. One is at the London Science Museum, one at the Chalk Pit Museum near Arundel, West Sussex, and the third outside the supermarket.

---

29. NEW INN BRIDGE,   HEW 1230
MERTHYR MAWR   SS 891 784

New Inn Bridge takes its name from a hostelry that once stood near this bridge which carried over the Ogmore river the main South Wales coaching route, now a minor road just south of the A48 road. The bridge is of masonry having four pointed arches. Cutwaters protect each pier and have stepped upper parts. Each span is about 15 ft and the width between the parapets is 10 ft.

The parapets contain openings through which sheep were once forced to leap into the deep pool beneath the bridge. Thus the bridge, which is several hundred years old and is a scheduled Ancient Monument, is also known as the Sheep Dip Bridge.

---

30. GLANRHYD BRIDGE   HEW 1084
                                                          SS 899 828

The Duffryn Llynfi tramroad linked ironworks and collieries in the vicinity of Maesteg to a small harbour at Porthcawl.

A branch from Tondu to Bridgend was carried across the Ogmore River by means of a three span masonry bridge known as Glanrhyd or Tyn y Garn bridge.

The bridge, on which a stone plaque records that it was built in 1829, is well proportioned and of good workmanship. It now carries a minor road which turns off the Bridgend-Maesteg road, the A4063, immediately to the north of Pen y Fai hospital. The bridge lies below a span of the Ogmore valley viaduct of the M4 motorway and served to support a section of the formwork during the construction of the viaduct. The two structures in such close proximity illustrate vividly the change in scale of transport provision over the last 150 years.

The tramroad was built to a gauge of 4 ft 7 in. and was intended for horse drawn traffic. In due course the lines were converted to standard gauge and from 1861 were worked by steam locomotives, although Glanrhyd bridge itself was never used by these.

Commemorative plaques now mark the termini at Maesteg, Bridgend and Porthcawl.

---

31. BARRY DOCKS HEW 1233
Illustrated below ST 124 668

In 1889 the Barry Railway Company built a dock between Barry Island and the mainland to capture a share of the export trade in South Wales coal, which had developed enormously over the

Barry Docks (Associated British Ports)

previous fifty years. The Engineer was John Wolfe Barry, one of whose assistants was I. K. Brunel's son Henry. T. A. Walker was the contractor.

The works comprised a basin 500 ft by 600 ft and a dock 3400 ft by 1100 ft divided by a mole into two arms at its west end. The wrought iron gates in the 80 ft entrance to the basin and between the basin and the dock were the first to be operated directly by hydraulic rams instead of through opening and closing chains.

At the north-east corner of the dock there was a dry dock 740 ft by 100 ft with a 60 ft entrance closed by a caisson.

The dock was liberally supplied with hydraulic hoists for tipping coal wagons into ships. The excavation generally was left with sloping sides. Walls were provided only around the basin, along the south side of the dock and at the positions of the hoists. Forty-six and a half feet high from dock bottom to coping, they were built in the dry, using massive limestone blocks.[17]

Breakwaters were constructed off the entrance, mainly with rubble from the excavations, faced with four ton stone blocks. On the end of the west breakwater is a cast iron lighthouse 30 ft high, 7 ft 9 in. in dia. at the base and 6 ft 6 in. dia. at the top (cf. HEW 1256).

Before the works started, cofferdams had to be formed by tipping across the channel to enclose the dock site and to enable the five million cubic yards of excavation to proceed. A Cornish engine of 267 000 gallons/h capacity was brought from Walker's Severn Tunnel works (HEW 232) to assist in the dewatering.

Crumlin Viaduct (Metius Chappell: *British Engineers*)

A pier was built on the Bristol Channel side of the docks, with a passenger line from Barry Island Station to the Pierhead Station through a tunnel under what is now Butlin's Holiday Camp.

A second dry dock, similar to the first, was opened in 1893.[18] The 34 acre No. 2 Dock was opened off No. 1 Dock in 1898; while in 1908 the Lady Windsor Lock, 647 ft long and 65 ft wide, and named after the wife of the Company Chairman, was opened directly into the sea to the west of the tidal basin.

By the end of the 19th century the coal trade had reached its peak and after a few years entered upon a continuous decline. To-day no coal is handled at the docks, but the present owners, Associated British Ports, have adapted them to handle a variety of trades, including the export of coke.

---

32.   MELINGRIFFITH WATER PUMP   HEW 392
                                                          ST 143 800

The Glamorganshire Canal, which was opened in 1798, drew water from a feeder which also supplied the Melingriffith Works in the Whitchurch area of Cardiff. The Canal Company was obliged to take its water from the tail-race below the Melingriffith Works, and this necessitated a pumping engine being installed to lift the water 12 ft into the canal feeder.

It was designed and constructed by John Rennie and William Jessop and operated for 135 years from 1807 until the canal's closure in 1942. The pump was well constructed (of American oak and cast iron) and survived until restoration work started in 1974.

The undershot water wheel is 18 ft 6 in. in dia. by 12 ft 6 in. wide and consists of three cast iron wheels mounted on an axle, originally of oak, now of steel. There are six cruciform cast iron spokes to each and the rims carry thirty paddles, 22 in. deep. The wheels drove pistons in two cylinders via timber rocking beams 22 ft long. The cylinders had a bore of 2 ft 8 in. and a 5 ft stroke.

There is a model in the Department of Industry, National Museum of Wales, Cathays Park, Cardiff.

---

33.   CRUMLIN VIADUCT   HEW 72
Illustrated on facing page   ST 213 986

Of all the many viaducts in South Wales, perhaps the most interesting and certainly the most dramatic was that at

Crumlin.[19,20] It is still worth describing although it was demolished in 1966. There is a small model in the National Museum of Wales and traces of the piers can be found on site.

Crumlin Viaduct had ten spans, divided into two groups, of seven and three, by a rock knoll. The total length approached 1700 ft. Crossing the valley of the Ebbw river at a height of 200 ft, the viaduct was built in 1857 by the Newport, Abergavenny and Hereford Railway for their extension to the Taff Vale. The piers

Newport Transporter Bridge

each consisted of 12 cast iron columns 12 in. dia. in three rows of four, with an additional raker at each end, covering an area of about 42 ft by 21 ft. Each column was in eight pieces with wrought iron cross-bracing to adjacent columns. There were four lines of Warren truss girders, each of 150 ft span and 15 ft deep, cross-braced by iron plates in 1868 and by steel from 1930. This was the first large scale multi-span use of this design in wrought iron.

The viaduct was designed and built by T. W. Kennard, the partner of James Warren, who devised the truss configuration which bears his name, and which was first used for a major bridge at Newark Dyke, on the Great Northern Railway main line (HEW 1023, SK 801 558).[21] The Crumlin columns were cast at his works and the wrought iron, supplied by Blaenavon Works, was fabricated at site and hoisted into position.[22]

---

34. FLIGHT OF LOCKS NEAR ROGERSTONE, MONMOUTHSHIRE CANAL

HEW 798
ST 286 884 to ST 279 886

The most spectacular set of locks in South Wales is at Rogerstone near Newport on the Crumlin branch of the Monmouthshire Canal, north of the M4 between junctions 26 and 27. There are fourteen locks in all, each with head and tail gates. The short lengths between adjacent locks are connected to side pounds and there are several header pounds between groups of locks in the flight. Thus water released on the downward passage of a boat could be partially re-used on later upward passages.

In all, the series of locks gave a rise of 168 ft in approximately half a mile of travel. Thomas Dadford, Jr was the Engineer for the Canal and the Crumlin branch was completed in 1798.

---

35. YNYS-Y-FRO RESERVOIR

HEW 120
ST 28 89

Near the upper end of the flight of 14 locks on the Monmouthshire Canal (HEW 798) is the Ynys-y-fro Reservoir. The year 1846 saw the establishment of a private company to supply water to the town of Newport, which then had a population of 19 000, and the passing of a Parliamentary Act which authorized the company to construct a reservoir at Ynys-y-fro and to lay pipes to convey water to the town. The Engineer was James Simpson.

The reservoir had a capacity of 71 million gallons and a maximum depth of 34 ft. It was contained by an earth dam with clay puddle core having side slopes of about 1 in 2½ and stone pitching on the submerged face.

In or about 1881 the Company commissioned a further reservoir upstream and adjacent to the original, the purpose apparently being to increase the storage capacity and to trap silt which was causing a discoloration of the water supply. The new dam was of similar construction to the earlier one; it now forms a causeway carrying a local road between the upper and lower reservoirs.

---

36.   NEWPORT TRANSPORTER BRIDGE   HEW 144
Illustrated on page 78   ST 318 862

Newport Transporter Bridge across the River Usk about 1½ miles south of Newport Town Centre was completed in 1906.[23] The only other similar bridges in Britain are at Middlesbrough (HEW 10, NZ 501 213)[24] and Warrington (HEW 140, SJ 597 877).[25] Located in an industrial area, the bridge is a slender, attractive structure. The design was chosen in order to give adequate headroom for the type of shipping using the River Usk at the turn of the century.

The bridge has a span, centre to centre of towers, of 645 ft and a clear headway from high-water level of 177 ft.

The towers are of open lattice steel construction each founded on four cast iron cylinders. There are sixteen suspension cables, four inside and four outside each of the stiffening girders, which are 16 ft deep and 26 ft 3 in. from centre to centre. The bottom booms are built-up plate girders, each lower flange carrying rails on which the travelling frame runs.

The platform car (33 ft by 40 ft wide) is hung from the travelling frame by thirty suspension ropes. The total weight of frame and car is just over 50 tons and they are moved by steel wire ropes wound on a drum worked by two electric motors, each of 35 BHP.

The engineers were F. Arnodin and R. H. Haynes and the builders Cleveland Bridge and Engineering Co. of Darlington.

---

37.   BIGSWEIR BRIDGE   HEW 361
                                                     SO 539 051

Travellers on the A466 between Chepstow and Monmouth temporarily cross from Wales into England over this bridge which

Chepstow Bridge

spans the River Wye in a beautiful scenic setting into which it blends with harmony. The bridge formed part of a toll road authorized in 1824 and was completed in 1828. The designer was C. Hollis of London.

The elegant cast iron arch of 164 ft span has four ribs of dumbbell section, in 16 segments with plain N spandrel bracing. It is supported on circular stone piers and is surmounted by a cast iron balustrade. The arch segments, cast at Merthyr Tydfil, are particularly well made.

Around the mid-19th century two masonry arched flood spans were added at each end, making the overall length 321 ft. The mock string course cast on the main arch at deck level is continued in masonry over the faces of the flood spans.

Despite its age and a carriageway only 12 ft wide the bridge copes with the heavy tourist traffic along the Wye Valley, with the aid of traffic lights and a 16 ton weight restriction.

---

38. CHEPSTOW BRIDGE   HEW 145
Illustrated above   ST 536 943

To the historically minded engineer, mention of Chepstow brings to mind Brunel's railway bridge; but just upstream, and not far from Chepstow Castle, is an older and interesting five span cast iron bridge carrying the A48 road over the Rive Wye, and opened in 1816 to replace the timber bridge shown in J. M. W. Turner's picture of the Castle.

The bridge was designed by J. U. Rastrick and cast by William Hazledine. The centre span is 112 ft, flanked by spans of 70 ft and 34 ft on each side. The spandrel fillings are of the radial grid pattern not unlike those of the Coalbrookdale Iron Bridge (HEW 136, Chapter 7) and some of Telford's bridges.

The graduated spans, the vertically curved road profile, the parapet fence and the decorative lamp standards combine to give the bridge its attractive appearance.

In spite of its age the bridge has continued to carry the traffic of a main trunk route by the expedient of traffic signals limiting the flow to one lane alternately in each direction. The centre span was strengthened in 1889 by the addition of steel ribs and the timber starlings have been improved by steel sheet piling, while extensive repairs were carried out in 1979 – 80.

CHAPTER 4

Avon and North Wiltshire

The area covered by this chapter is dominated by the City and Port of Bristol and its communications with London, the Midlands, South Wales and the South-West Peninsula. The City's industries grew up on locally-produced raw materials and on imports of foreign goods, including sugar and tobacco from the Americas.

The Bristol coalfields extended in a wide arc from south-west to north-east, passing round the east of the City, while further afield the Somerset coalfield, south-east of Bristol and south-west of Bath, was worked until 1975.

To the south the Mendips once provided lead, worked by the Romans, and zinc for brassmaking, and still produce limestone in large quantities. The Mendip area is an important source of water supply. Bath is largely built of Bath Stone, still mined east of that City.

Iron ore was found locally, and iron was also brought down the Severn from the Forest of Dean.

From the 18th century, expanding trade led to the construction of canals, the establishment of turnpike roads (these last not uninfluenced by the growth of Bath as a spa) and at Bristol the building and improvement of docks, continued up to the present day. In the 19th century came the railways, pioneered by I. K. Brunel's Great Western Railway to connect Bristol with London.

It is interesting to compare the development of communications from London to Bristol and beyond, across Wales to Ireland and the Atlantic, with similar developments in North-West England and North Wales. Today the area is traversed from east to west by the M4 motorway from London to South Wales, which intersects the M5 motorway from South-West England to the Midlands on the northern outskirts of Bristol.

Many of the original local industries and the facilities which served them have disappeared or fallen into disuse, to be replaced by developments using modern technology. It is interesting to note, however, that some of the ancient industries have continued

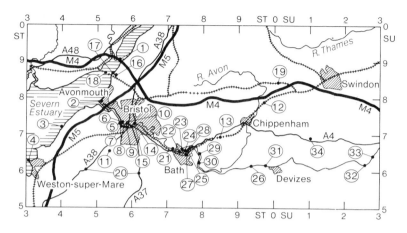

1. Severn Road Bridge
2. No. 1 Granary, Avonmouth Docks, Bristol
3. Clevedon Pier
4. Birnbeck Pier, Weston-super-Mare
5. Clifton Suspension Bridge
6. Bristol City Docks
7. Brunel Swing Bridge, Cumberland Basin, Bristol
8. 'B' Tobacco Warehouse, City Docks, Bristol
9. SS *Great Britain*
10. Bush House, Bristol
11. Bristol Waterworks
12. Great Western Railway, Swindon – Bristol
13. Box Tunnel
14. Temple Meads (old) Station, Bristol
15. Temple Meads (present) Station, Bristol
16. Bristol and South Wales Union Railway

17. The Binn Wall
18. Severn Tunnel
19. South Wales and Bristol Direct Line
20. The Bristol and Bath Turnpike Roads
21. New Bridge, Bath
22. Victoria Suspension Bridge, Bath
23. Green Park Station, Bath
24. Bath Roman Water Supply and Drainage
25. Pulteney Weir, Bath
26. Kennet and Avon Canal
27. Widcombe Locks, Bath
28. Cast iron bridges in Sydney Gardens, Bath
29. Claverton Pumping Station
30. Dundas Aqueduct
31. Devizes Locks
32. Crofton Pumping Station
33. Great Bedwyn Skew Bridge
34. Silbury Hill

Outstanding British suspension bridges

Bridge	Date	Span: ft	HEW number
Union[4]	1820	361	143
Menai	1826	579	109
Clifton	1836–64	702	129
Tamar[5]	1961	1100	202
Forth[6]	1964	3300	203
Severn	1966	3240	201
Humber	1981	4626	204

into the present century, such as the smelting of zinc at Avonmouth from ores imported through the docks, while lead shot continues to be cast by traditional methods in a modern shot tower in
the heart of Bristol. The city remains a principal centre of the
tobacco trade.

Of all the past developments much remains as a reminder of our
civil engineering heritage and of the engineers themselves such as
Jessop, Rennie and Brunel: for instance Rennie's Kennet and
Avon Canal built to link the Severn and Thames and paralleled by
Brunel's Great Western Railway.

---

1. SEVERN ROAD BRIDGE   HEW 201
ST 560 901

The traveller by road will cross from South Wales to England on
the M4 motorway which is carried across the Wye and Severn on
two major bridges, built between 1961 and 1966. They were opened
by HM Queen Elizabeth II on 8 September of the latter year.

The Wye Bridge is a stayed cantilever bridge with a 770 ft main
span and two 285 ft side spans. It is joined to the Severn Suspension bridge[1-3] by a ten-span steel viaduct across the Beachley
Peninsula.

The main span of the Severn Bridge is 3240 ft with 1000 ft side
spans. The road is 120 ft above the water. The steel towers are
400 ft high and 104 ft 6 in. wide, and the main cables are 20 in.
dia.

The bridge is rather young to be regarded as an Historical
Engineering Work, but although it has only the third longest span
in Britain (and the eighth in the world) it is important in terms of
technical development. It may be compared with other outstanding British suspension bridges (see table on facing page).

The design of the bridge pioneered the arrangement of the
hangers in a triangular pattern while the deck is an integral
streamlined box girder instead of the old type trussed stiffening
side girders with separate cross girders and decking. The Humber
Bridge and the Bosporus bridge in Turkey follow the same pattern.

The substructure was designed by Mott, Hay and Anderson
and built by John Howard; the superstructure by Freeman Fox
and Partners and Associated British Bridge Builders (Arrol,
Cleveland Bridge and Dorman Long).

The numbers and weights of heavy vehicles on the roads are now
much greater than were envisaged when the bridge was designed.

In 1982 new design loadings and standards for bridges were introduced, dramatically higher than those of twenty years earlier.[7] Within the profession there were differences of opinion as to whether these should be applied retrospectively to the Severn Bridge or whether all that was necessary was to deal with the teething troubles experienced by the hangers and with other defects that had developed. In the event, the Department of Transport decided to play safe and to carry out comprehensive strengthening of much of the structure.

Immediately downstream of the Severn Bridge is the longest overhead cable crossing in Britain (HEW 1146, ST 556 891).

No. 1 Granary, Avonmouth Docks

From the bridge can be seen, a few miles upstream on the east bank of the Severn, Oldbury Power Station (HEW 1180, ST 606 943), the first nuclear power station in Britain to have a concrete pressure vessel.

---

2. NO. 1 GRANARY, AVONMOUTH DOCKS, BRISTOL Illustrated on facing page	**HEW 1012** ST 513 785

The M5 motorway south from Bristol crosses the River Avon close to the Avonmouth Docks of the Port of Bristol Authority.

Number 1 Granary, the earliest of the five reinforced concrete granaries in the Royal Edward Dock, was built at the same time as the dock in 1908 and formed part of the first bulk grain handling installation in the Port.

The building is 220 ft long, 71 ft wide and some 85 ft high. The north end housed intake elevators and weighers; the middle is occupied by 65 rectangular storage silos; and at the south end there was floor storage on six floors, later given over to additional handling equipment.

Like Weaver's Mill at Swansea (HEW 354, Chapter 3), the building was designed to the Hennebique system by L. G. Mouchel. It remained in full use until 1983, when it was taken out of service following the decline in the grain trade. The builders were John Aird and Co., the main contractors for the Royal Edward Dock.

Nearby are the contemporary O and P Sheds (HEW 994, ST 511 784 and ST 512 786), also designed by Mouchel and still in full use.

Adjoining the Royal Edward Dock[8] is the Avonmouth Dock (1878),[9] while across the Avon are Portishead Dock (1875) and the Royal Portbury Dock opened in 1977,[10] whose entrance lock (HEW 894, ST 497 780) is at 1204 ft the longest in Britain.

---

3. **CLEVEDON PIER**	**HEW 430** ST 401 719

Further down the coast is Clevedon Pier, probably the most elegant of the seaside piers around the British coast.[11]

Eight 100 ft spans comprise a 16 ft 6 in. wide pier leading out to a pierhead carrying three elegant cast iron pavilions.

The four legs of each supporting trestle are made of pairs of Barlow rails, of 'top-hat' section, redundant from the South Wales Railway (HEW 1199, Chapter 3), rivetted foot-to-foot. Each leg is spigotted into a 2 ft dia. cast iron screw pile. Beneath the deck one rail of each leg curves away from its twin to form, with a corresponding rail of the next trestle, a semi-circular bracing which, with the assistance of a series of struts, supports the main deck girder. The simplicity and slenderness of the construction render the pier most pleasing visually.

Designed by J. W. Grover[12] and R. Ward, with the ironwork fabricated by the Hamilton Ironworks, Liverpool, the pier was opened in 1869.[13]

The present pierhead was built in 1893 at an angle to suit the run of the tide, and to replace the original pierhead which was square to the pier and therefore inconvenient for the berthing of ships.

In the heyday of pleasure steamers, the pier served many trips between Bristol and the resorts in the Bristol Channel as well as affording an amenity for holiday makers and anglers.

Latterly, changing public taste in leisure activities greatly reduced the use of the pier and consequently the revenue, with adverse effects on its maintenance. Unfortunately in 1970 two of the outer spans collapsed, leaving the pierhead inaccessible.

The pier is one of only three seaside piers listed for preservation, the others being the Garth Pier at Bangor (HEW 427, SH 584 732) and the West Pier, Brighton (HEW 212, TQ 303 041).

Since 1980 the pier has been in the hands of a Preservation Trust which, with limited financial resources, set about a modest programme of rehabilitation. In March 1984 the National Heritage Memorial Fund and the Historic Buildings Council each made a grant of £500 000 to the Trust towards complete restoration, which began in the spring of 1985, and the future of the pier is now assured.

| 4. | **BIRNBECK PIER,** | **HEW 434** |
| | **WESTON-SUPER-MARE** | ST 307 626 |

Birnbeck Pier is one of the few remaining piers designed by Eugenius Birch, who was responsible for no fewer than fourteen seaside piers in Britain.

The pier is unusual in that the 'pierhead' is in fact the small Birnbeck Island, which was levelled and stepped to form a promenading area.

The pier was built by the Isca Iron Co. of Newport (Gwent) and was opened in June 1867.[14]

Iron girderwork 1150 ft long, with a 20 ft wide timber deck and supported on fifteen clusters of eight piles, connects the island to the shore. A further 250 ft of pier structure extends northward from the island to form a landing stage. The island also accommodates the house and launching slip for the Weston-super-Mare lifeboat.

The fencing to the pier incorporates ornamental lamps and decorative seating, with small projecting bays at intervals, and with the slender and economical iron work of the structure, lends the pier an air of unmistakeable Victorian charm.

Like Clevedon Pier, Birnbeck catered for the once-thriving pleasure steamer trade in the Bristol Channel, but also like Clevedon, now proves of little attraction to the public.

During the Second World War the pier, as *HMS Birnbeck*, accommodated a small naval establishment.

In February 1984, a pipe-carrying pontoon broke loose from beach replenishment work further up the estuary, drifted down on to the pier, broke off two diagonally opposite legs of the third pile cluster from the island and cracked two other legs, but without causing any collapse of the structure.

SS *Great Britain* passing under Clifton Suspension Bridge (South West Picture Agency Ltd)

5. CLIFTON SUSPENSION BRIDGE   HEW 129
Illustrated on page 89   ST 565 731

The A4 trunk road from Avonmouth to Bristol passes under the Clifton Suspension Bridge.

In 1753 one William Vick left £1000 to be invested until £10 000 had accumulated, when the money was to be applied to realizing his dream of a bridge across the Clifton Gorge.

In 1829 the trustees advertised a design competition, which after a fierce and controversial contest was won by I. K. Brunel, then aged 25, and he was appointed Engineer.[15,16]

Work began in 1836 but by 1842 the money ran out when only the towers and anchor tunnels had been constructed. The chains, which had already been made by the Copperhouse Foundry, Hayle, were sold to form part of Brunel's Royal Albert Bridge at Saltash (HEW 29, SX 435 587).

After Brunel's death in 1859, a number of members of the Institution of Civil Engineers decided to complete the bridge as a memorial to him. Work began in 1860 and was completed in 1864 to amended designs by John Hawkshaw and W. H. Barlow.[17]

Bristol City Docks circa 1947 (Port of Bristol Authority)

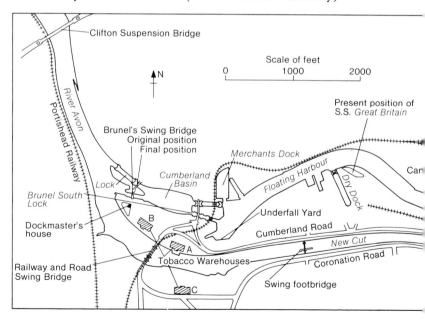

At 702 ft 3 in., the bridge for a time had the longest span in the world. Its 31 ft wide deck is 240 ft above river level. The suspension chains are made up of 7 in. × 1 in. wrought iron links, 24 ft long and fittingly, some of them came from the Hungerford footbridge across the Thames in London, the only other suspension bridge designed by Brunel. Completed in 1845, it was demolished to make way for the Charing Cross Railway Bridge.

6. BRISTOL CITY DOCKS   HEW 861
Illustrated below   ST 567 723 to ST 616 726

Within sight of the suspension bridge is the entrance to the City Docks which traverse Bristol from west to east.

Their history began in 1247 with the construction of the Broad Quay in the River Frome. From then only relatively minor works were carried out until the increasing size of ships in the 18th century led to the forming, between 1804 and 1809, of the Floating Harbour, one of the greatest civil engineering undertakings of its time.

This dock system, designed by William Jessop,[18] comprised parts of the courses of the Rivers Avon and Frome, made possible

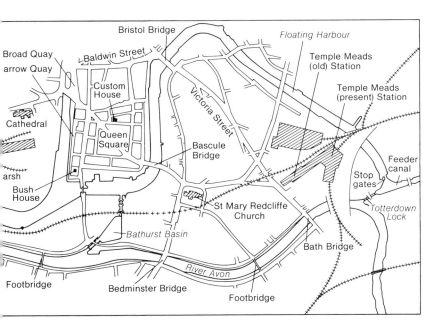

by the digging of a new channel for the Avon (the New Cut) from
Totterdown to Rownham, where there were two entrance locks
from the Avon into the basin entrance, which was joined to the
harbour by a Junction Lock on its south side. With a dam across
the Avon at Netham, water was fed to the harbour by a feeder
canal from Netham to Totterdown. Barge locks (long since closed)
communicated with the New Cut at Totterdown and Bathurst
Basin.

In 1832, on the advice of I. K. Brunel, Netham Dam was raised,
scouring sluices were made into the New Cut from the west end of
the harbour and in 1843 a drag-boat to Brunel's design (HEW
1163), which continued in use until 1961, was made for moving silt
from the dock walls.

Brunel was Engineer for a new and larger South Lock built in
1849, which had the first ever wrought iron buoyant gates (HEW
1173).[19]

In 1873 the present entrance was built on the site of the North
Lock with a new Junction Lock on the north side of the
Cumberland Basin. From then until the 1960s, various relatively
minor but important alterations and improvements were made to
the system.

The docks closed to general shipping in 1974. The City is
developing the area as a public amenity with strong emphasis on
preserving the many surviving examples of its engineering past.

---

7. BRUNEL SWING BRIDGE,   HEW 926
CUMBERLAND BASIN, BRISTOL   ST 568 724

This interesting relic of Brunel's early girder work is still to be
seen, swung parallel to the entrance lock in the shadow of the large
1965 swing bridge over the lock. Originally 120 ft 9 in. long, it has
wrought iron main girders with tubular flanges, the top circular,
the bottom triangular. Twin iron tie-bars run from end to end
within the top flange.

The history is a little complex. The bridge was designed and
built for Brunel's South Lock (HEW 1173, ST 567 723) in 1849. A
similar bridge was built over the North Lock in 1863. When this
lock was reconstructed in 1874 its bridge, now too short, was
transferred to Bathurst Basin (and subsequently replaced). The
Brunel bridge was shortened by 10 ft and transferred to the North
Lock. A copy of it was built in 1875 and used at the South Lock
where it is still in use but fixed, as that lock is now disused.

8. 'B' TOBACCO WAREHOUSE,   HEW 1011
CITY DOCKS, BRISTOL   ST 568 722

'B' Tobacco Warehouse is one of three imposing multi-storey brick-clad warehouses almost identical in appearance, near the Cumberland Basin entrance to Bristol City Docks.

'A' Tobacco Warehouse, on the north bank of the Avon (ST 570 721), was built between 1903 and 1905 as a steel-framed structure with floors of steel joists and concrete jack-arches, founded on concrete footings as much as 50 ft below ground.

When the construction of No. 1 Granary at Avonmouth Docks (HEW 1012) went ahead in reinforced concrete, it was decided to use that material for 'B' Warehouse.

'B' Warehouse was built between 1906 and 1908 by William Cowlin and Sons of Bristol, who also built warehouses A and C, the latter (ST 569 720) on the south bank of the river in 1915, also in reinforced concrete.

The designer, Edmond Coignet, was the son of François Coignet, a pioneer of structural concrete in France.

All three buildings are 214 ft long, 104 ft wide and 93 ft 9 in. high to the top of the parapet. The ground floors are 15 ft 6 in. high, the remaining eight floors 8 ft 6 in. to suit the stacking of cylindrical 'tierces' of leaf tobacco two-high. The roofs are of north light steel trusses clad in slates and patent glazing.

Tobacco Warehouse 'B' is founded on reinforced concrete piles, Warehouse 'C' directly on rock near the surface.

With the changes in the handling of leaf tobacco in the 1960s the warehouses ceased to fulfil their original function but are still in use for general warehousing.

9. S.S. *GREAT BRITAIN*   HEW 247
Illustrated on page 89   ST 578 723

Further up the docks Brunel's *Great Britain* lies in the former Great Western Dry Dock, built for her construction and in which she is now being restored to her original form as a memorial to one of the most historic ships ever built and to her great designer, whose versatility she exemplifies.[20,21]

She was the largest ship of the time with a displacement of 3675 tons and the first with iron hull and screw propulsion. She carried 252 passengers, 130 crew, 1200 tons of cargo and 1200 tons of coal.

Her engines, using steam at 5 lb/sq.in., developed 1600 hp at 18 rev./min. The six-bladed propellor, 15 ft 6 in. in dia. was driven at 54 rev./min. through four sets of chain drives. On trials the ship achieved a speed of 11 knots under steam alone. Sails, on six masts, were provided to assist propulsion when the wind was favourable.

Launched on 19 July 1843 by the Prince Consort, she made four voyages between Liverpool and New York before running aground in Dundrum Bay in 1846. After an ingenious salvage operation by Brunel's nautical adviser, Captain Claxton, in August 1847, she was sold to new owners in 1850. They changed the propellor to three-bladed, installed new boilers at 10 lb/sq.in. and increased the passenger accommodation to 730. In 1852 the ship began sailing between Liverpool and Australia.

Up to 1876 she made 32 such voyages plus two to New York and had been requisitioned as a troop carrier in the Indian Mutiny and the Crimean War, 1853 – 56 (in which Brunel was also concerned as the designer of prefabricated timber hospital buildings).

In 1882 she was sold and converted for carrying coal under sail between South Wales and San Francisco, but in May 1886 she was damaged off Cape Horn and put into Port Stanley, Falkland Islands. There she served as a coal and timber store ship until in 1937 she was deliberately holed and beached.

In 1970 the S.S. *Great Britain* Project, in a remarkable salvage operation, raised the ship onto a pontoon on which she was towed back to Bristol. Released from the pontoon in the Avonmouth dry dock, she was towed up the Avon and on 19 July 1970 was berthed in the Great Western dry dock.

---

10.   BUSH HOUSE, BRISTOL   HEW 624
                                                    ST 587 724

Not far from the *Great Britain* is Bush House, an excellent example of the adaptation of an old building of historical and architectural importance to modern use without departing unduly from its original appearance and setting.

The building, originally of five storeys, stands in a prominent position at the junction of two arms of the Bristol City Docks. Long known as Bush's Warehouse, it was built between 1832 and 1837 and was one of the finest examples of its kind and time in the City. With the disappearance of general cargo trade from the City Docks in the late 1960s it fell into disuse. Between 1974 and 1976 the present leaseholders completely reconstructed the building but retained the original shell.

Internally the original timber floors, carried on cast iron columns, were replaced by a piled reinforced concrete structure independent of the rest of the building. A steel framed mansard roof was added to give a sixth storey above original roof level and, if anything, this enhances the original appearance.

The walls of dark grey pennant stone with Bath stone details were cleaned and repaired. The disposition of doors and windows was retained. On the south elevation, which is the most symmetrical, with the most striking grouping of details, even a small wall crane was retained.[22]

The two lowest storeys now house a privately owned arts and leisure complex; the rest are offices.

The building was depicted on the 28p postage stamp which was part of a set issued on 10 April 1984, featuring urban renewal and commemorating the 150th anniversary of the Royal Institute of British Architects.

11. BRISTOL WATERWORKS   HEW 1242
Illustrated on page 96

The population of Britain nearly doubled during the second half of the Industrial Revolution. Towns multiplied their size and became congested. Water supply and sewage disposal, never good, became totally inadequate. The 1842 Report *The Sanitary Conditions of the Labouring Classes in Great Britain* triggered off a great burst of sewerage schemes which in turn increased the demand for water, for domestic use as well as for industry.[23] Drainage into streams and rivers led to pollution and new sources of supply had to be sought. Many of the new developments took place in the same period as the great expansion of railways and, like them, placed considerable demands on the civil engineering resources of the nation.

In 1844 Bristol was described as having the worst water supply of any large town. In 1846 Bristol Waterworks Company was incorporated and appointed James Simpson as Consulting Engineer.[24] He recommended bringing in the principal supply from springs in the Mendips south of the City, where at a height of 400 ft advantage could be taken of a 50 square mile catchment area with an annual rainfall of 41 in.[25]

The works involved collecting spring water from the Mendips in the vicinity of Chewton Mendip (ST 55 50) and conveying it in an aqueduct (the Line of Works) to a reservoir at Barrow (ST 53 67), on an average gradient of five feet in a mile. There are four miles of 30 in. dia. pipe and four miles of tunnel. On the route of the tunnel there are remarkable crossings of valleys at Winford (ST 54 65), Leigh and Harptree by wrought iron rivetted tubes, the first two 825 ft long, the third 350 ft. The tubes are egg-shaped, 4 ft 7½ in. high, 3 ft 6 in. wide, supported on cast iron saddles on the tops of masonry piers at 50 ft centres, as much as 60 ft high, with cast iron ball bearings to allow longitudinal movement. The aqueduct is still in service today and forms an integral part of the Mendip supply system.[26]

From Barrow the water was piped to service reservoirs at Clifton, Durdham Down and Bedminster, then on the outskirts of the City. The work was completed in 1851 and one of the original mains is still in service. After 130 years service, this 20 in. dia. cast iron main was found to be badly corroded on the outside, but in fair condition internally. In 1983 a start was made with lining the main with continuously welded plastic pipe pulled through from excavations made at intervals. This was the first application in this

Bristol water supply: pipeline at Winford

area of this relatively new technique, which avoids the wholesale digging up of pipelines for renewal.

The initial works were designed for four million gallons per day. In 1866 a further storage reservoir was completed at Barrow and a third was added in the period 1897–1900. Between 1867 and 1870 a further six million gallons per day were tapped from springs at Chelvey (ST 47 68) and pumped to Barrow. Simpson's 60 hp pumping engines at Chelvey were scrapped in 1937.

In the period 1898–1901 a 700 yd long earth dam was built across the Yeo valley near Blagdon (ST 50 60) which provided a 470 acre reservoir with 1700 million gallons capacity.

In 1930 springs in the Cheddar Gorge were tapped for pumping to Barrow, and in 1937 Cheddar Reservoir (ST 44 53) was completed to store the excess water for use during dry periods.

In 1956 HM Queen Elizabeth II opened Chew Valley Lake, at the time the largest man-made lake in Britain, and still one of the six largest. An earth dam (ST 57 61), nearly a quarter of a mile long, with a concrete filled cut-off trench into the underlying marl, impounds 4500 million gallons in a lake $2\frac{1}{4}$ miles long and $1\frac{1}{2}$ miles wide. This serves not only Bristol but also as far afield as Bath, Radstock and South-West Gloucestershire.

Blagdon and Chew Valley Lake, with the Barrow reservoirs, are important trout fisheries, their trout hatchery having a world-wide reputation. Cheddar and Chew offer small boat sailing.

The company also extracts Severn water from the Gloucester and Sharpness Canal (HEW 466, Chapter 5), with treatment works at Purton (SO 69 04) and Littleton (ST 60 89).

---

12. **GREAT WESTERN RAILWAY,**	**HEW 1071**
SWINDON – BRISTOL	SU 149 856 to ST 597 724

The Great Western Railway between Bristol and London, authorized under an Act of 31 August 1835, was I. K. Brunel's first venture into the field of railway engineering, in which he was to become a master.

From London to the summit of the line at Swindon and a little way beyond, gradients were easy and few major engineering works were necessary other than at river crossings, but over the next 41 miles west of Swindon the terrain was more formidable and numerous major works were required. This suggested that higher speeds might be attained on the London side, using locomotives with larger diameter driving wheels than those used westwards.

A pause to change engines and allow the passengers time for refreshment was not unreasonable in 1840, when the newfangled railway was replacing horse-drawn stage coaches operating at about a quarter of the speed and with a similar fraction of the distance between stops.

Swindon was even then much more at the hub of transport systems than might be supposed.

It is only five miles north of one of the very oldest land routes in the UK, the Great Ridgeway, which is crossed by the very important Roman road from Portchester to Cirencester via Winchester – now the A345 leading to the A419 (Ermin Way) – at SU 193 794, about six miles north of the Bath road (now the A4) near Savernake. Junction 15 on the M4 motorway is on the same Roman road, almost in Swindon itself.

Some 12 miles to the north-west is Cirencester where the Ermin Way joins Foss Way, the major Roman artery to Lincoln and the north.

The Kennet and Avon Canal (HEW 1034) is not far south of the Bath Road, and had only been opened in 1810, as was the Wilts and Berks Canal, also running east-west, this time through Swindon itself. It was closed in 1906 and abandoned in 1914.

To the north again, between Swindon and Cirencester, are the headwaters of the Thames and the Thames – Severn Canal opened in 1789 (HEW 1139, Chapter 5). Linking these two, between Swindon and Latton, near Cricklade, was the North Wilts Canal, opened in 1819 but now long since abandoned. All these waterways benefitted for a few years from the work of constructing the railways and supplying their needs.

At the outset, in 1840, on the advice of Daniel Gooch, Brunel's young Locomotive Superintendent appointed at the age of 25, the GWR established a locomotive depot or running shed at Swindon to cater for 48 locomotives under cover and to provide repair facilities for a further 54.

This developed into Swindon Works (SU 125 835) – one of the largest railway establishments in the world for construction and repair of locomotives, carriages and wagons. It eventually occupied about 320 acres, of which over 70 acres were roofed. At its peak there was a staff of 12 000 and a technical and manufacturing capability almost unrivalled in its day.

Relics connected with I. K. Brunel may be seen at the GWR Museum at Swindon.

Between Swindon and Chippenham, deep cuttings and high embankments were required over poor ground.

From Chippenham to Bath were the hardest 13 miles of the line. Starting with the viaduct at Chippenham (HEW 380, ST 912 731) there were long embankments and deep cuttings in rock; then Box Tunnel (HEW 236); followed by Middle Hill Tunnel (HEW 381, ST 820 687) and a bridge across the River Avon at Bathford (ST 785 671). At Bath the Avon was crossed just east of the station by an 88 ft span stone arch (rebuilt 1926–27). Two 80 ft laminated timber arches,[27] replaced by an iron girder bridge in 1879, crossed the river west of the station. On both sides of the Avon the bridges are approached by long stone viaducts.

Between Keynsham and Bristol a number of short tunnels were needed to take the railway along the steep hillside above the River Avon.

Outside the Bristol terminus three bridges were needed to cross waterways. Two have been replaced by modern steel structures. The survivor, Brunel's 'Gothic' masonry arch across the Avon (HEW 1108, ST 614 724) is itself masked from view by modern bridges on each side.[28]

The line was opened throughout from London to Bristol on 30 June 1841, following completion of Box Tunnel.

Brunel had adopted the broad gauge of 7 ft $0\frac{1}{4}$ in. on the grounds that this would permit larger locomotives, higher speeds and greater comfort and safety for the passengers. But the 'standard' gauge of 4 ft $8\frac{1}{2}$ in. had been firmly established on the other railways which had spread throughout the country since the Stockton and Darlington Railway (HEW 85) was opened in 1825,

Box Tunnel: west portal (Bourne)

Temple Meads (old) Station (Bourne)

and in 1846 the Gauge Act came down firmly on the side of standard gauge for further railways.

Nevertheless the GWR broad gauge penetrated to the extremity of the South-Western peninsula and into South Wales, though, quite early on, a third rail was added to various sections of track in order to permit through running with other lines. London to Bristol had mixed gauge by 1875. In 1869, by which time there were 1500 miles of broad gauge track, the Company had conceded that the broad gauge had lost the contest and embarked on a programme of conversion, which was not completed until 1892.

13. BOX TUNNEL HEW 236
Illustrated on page 99 ST 830 689 to ST 857 694

Box Tunnel, some five miles east of Bath, is the most important of the substantial works necessary to bring Brunel's Great Western Railway down to Bristol from the summit level at Swindon.

One mile and 1452 yd long, with a downward gradient of 1 in 100 towards Bath, it was then by nearly 800 yd the longest railway tunnel so far built. It is 30 ft wide, which accommodated two of Brunel's broad gauge tracks.

Work began in 1836 with the sinking of seven shafts,[29] the deepest 290 ft, up which the excavated material was raised by horses at the surface operating hoisting drums.

Two contracts were let for the driving of the tunnel.

In the contract for the westernmost half mile, which was in rock and unlined, excavation was by blasting with gunpowder, of which a ton was used each week. Considerable ingress of water was dealt with by two 50 hp pumps.

In the remainder, in softer strata, excavation was by pick and shovel, using techniques which had been developed over many years in the construction of canal tunnels, and the tunnel was lined with brick. Brunel had, of course, already gained experience in this class of work in helping his father on the Thames Tunnel.

Work was completed in 1841. At the peak 4000 men and 300 horses were employed. The deaths of 100 men during the progress of the works underline the hazards of construction.

The monumental west portal is a notable feature.

| 14. TEMPLE MEADS (OLD) STATION BRISTOL | HEW 224 ST 596 724 |

Illustrated on facing page

The original terminus of the Great Western Railway was opened in 1841. No longer in use as a station it served for a number of years as a covered car park. A trust has been formed and with the backing of British Rail has restored the building as a Brunel Engineering Centre.

Apart from the neo-Tudor frontage on Victoria Street the station is notable for its timber roof, the centre span of which, at 74 ft, is wider than the roof of Westminster Hall.[30] The roof is basically a series of cantilevers, supported on lines of cast iron columns just inside the platform edges, meeting at the ridge, and tied down at their rear ends to the outside walls of the building. Brunel evidently felt that such simple construction on its own was not worthy of the western terminus of his great undertaking, the Great Western Railway, and added further embellishments in the shape of curved brackets, with lattice infilling, at the springings of the cantilevers, together with ornamental timber pendants to give the appearance of a hammerbeam roof.[31]

| 15. TEMPLE MEADS (PRESENT) STATION, BRISTOL | HEW 436 ST 598 725 |

The main Great Western Railway line from London joined the Bristol and Gloucester Railway (HEW 1075, Chapter 5) on an east-west alignment before swinging south-west over the Floating Harbour into the original Brunel terminus (HEW 224). Adjacent to this, but at right angles, was the terminus of the Bristol and Exeter Railway (HEW 1072), whose line set off southwards across the New Cut and whose 1854 Jacobean style offices can be seen on the south-east side of the present station approach.

By 1865 enlargement and alterations became necessary and in 1876 a new station roof was built over a 970 ft radius curve linking all routes as part of the Great Western and Midland Joint Station.

A rather similar modernization took place at York at much the same time and it is interesting to compare the two roof designs. Whereas York (HEW 239, SE 596 517)[32] has four spans of wrought iron plate girders without ties, Bristol has one 125 ft span consisting of 26 lattice ribs at 18 ft 9 in. centres, trussed with $3\frac{1}{2}$ in. and 3 in. dia. main ties and 4 in. dia. tube struts.

The station facade, at the head of a wide approach ramp, is in an ornate Gothic style.

16.	**BRISTOL AND SOUTH**	**HEW 1033**
	WALES UNION RAILWAY	ST 609 726 to ST 504 882

The Bristol and South Wales Union Railway represented one of the steps in reducing the distance by rail to South Wales.[33] Opened in 1863[34] for passengers only, eleven miles of broad gauge track shortened the distance from Bristol to Cardiff from 94 miles to 38 as compared with the previous route via Gloucester. The Engineer was R. P. Brereton, I. K. Brunel's principal assistant.

The line led from a junction with the Great Western Railway east of Temple Meads, via a deep cutting at Horfield (ST 607 776) and 1245 yd of tunnel at Almondsbury (ST 598 824) to New Passage (ST 543 863) on the east bank of the Severn. Here the trains ran along a timber jetty 1635 ft long. A steam ferry made the two mile crossing to a similar jetty, 708 ft long, at Black Rock (ST 514 881), whence about a mile of railway connected with the Gloucester – Newport line at Portskewett. The jetty heads incorporated stairs and ramps and floating pontoons. Marc Brunel, the father of Isambard, is reputed to have been the first to suggest the idea of floating pontoons as landing stages on a visit to Liverpool in 1826. When the Severn Tunnel was opened in 1886 the ferry was abandoned but much of the railway work on the Bristol side was incorporated in the new route.

At Black Rock a few pile stumps remain and just back from the foreshore is a masonry arch carrying a footpath over the cutting that led to the jetty. At New Passage the line of the jetty can just be traced in the foreshore mud and a short length of the masonry abutment is built into the sea wall (HEW 1207).

---

17. THE BINN WALL   **HEW 1207**
                                         ST 539 860 to ST 545 865

Extending from Severn Beach to just north of the site of the Bristol and South Wales Union Railway jetty at New Passage (HEW 1033), the Binn Wall forms part of the sea defences of the low-lying land between Avonmouth and Aust and dates from at latest the first half of the 17th century.[35]

Directly fronting the Severn Estuary, with no intervening saltings, the wall is exposed to the full force of the sea. Between 1816 and 1818, following storm damage, the greater part was rebuilt as an earth bank 10 to 17 ft high, faced on the seaward side with $3\frac{1}{2}$ ft thickness of stone pitching. At ST 540 858 a stone slab let into the face of the wall, and bearing the date 1818, records the names of the Surveyor and Mason. In 1979 the wall, which had been overtopped in storms a few years earlier, was raised and widened, and at the north end was reconstructed with its seaward face protected by interlocking concrete blocks and retained by steel sheet piling; a fine example of modern sea defence techniques.

Immediately beyond the north end of the wall is a natural bank of gravel. In 1823 J. L. McAdam, Surveyor to the Bristol Turnpike Trust (HEW 1061), was fined by the Wall Commissioners for removing gravel for use on his roads and later the B&SWU Railway had to be restrained from using this material.

---

18. SEVERN TUNNEL   **HEW 232**
                                         ST 480 876 to ST 545 854

The construction of the Severn Tunnel, the longest railway tunnel in Great Britain (4 miles 624 yd) further shortened the route between London and South Wales and is the story of a battle against nature in the shape of groundwater and the tides, well described in the classic book by T. A. Walker the contractor.[36]

The tunnel crosses the Severn between Pilning in Avon and Sudbrook in Gwent, where the estuary is $2\frac{1}{4}$ miles wide and the

Bristol and Bath Turnpike Roads: signpost; milestone

range of the tide can be 50 ft. Although a great deal of the rocky bed is uncovered at low tide the main channel is 80 ft deep below general bed level. The Engineer was Sir John Hawkshaw.

The Great Western Railway began the work in 1873 with the sinking of a shaft at Sudbrook and the driving of a drainage heading towards the river. Four and a half years later only the shaft and 1600 yd of heading had been built. In 1877 contracts were let for additional shafts on both sides and headings on the line of the tunnel. In October 1879, when the headings had nearly joined up, ingress of water from an underground river, the Great Spring, inundated the original Sudbrook workings. A contract was then let to Walker for the completion of the whole works, but it was not until January 1881 that the spring was walled off and the works de-watered.

In April 1881 water broke into the Gloucestershire workings from a hole in the river bed. This hole was sealed with clay in bags and with concrete. In October 1883 the Great Spring broke in again and a week later a spring tide flooded the long deep cuttings on both sides. A heading from the Sudbrook shaft was driven to tap the Great Spring. With the water so diverted, the 300 yd of tunnel in which the Spring had been walled was completed.

Two 50 in. Bull engines, made by Harvey of Hayle, and installed for dealing with the flooding, still survive, one in the Science Museum, London, one stored by the National Museum of Wales, Cardiff.

The brick lining was completed for the full length of the tunnel in April 1885, and in September a special train took through the tunnel a party which included Sir Daniel Gooch, now Chairman of the Company.

In 1886 a shaft was sunk beside the tunnel and six pumps with 70 in. beam engines, also by Harvey of Hayle, were installed to deal with the twenty million gallons of water per day, mainly from the Great Spring, at the Sudbrook pumping station. Three of these engines remained in use until the station was electrified in 1961. The pumping station supplies water to a nearby paper mill.

The tunnel was opened to traffic on 1 December 1886.

---

19.   SOUTH WALES DIRECT LINE   HEW 1073
SU 067 818 to ST 611 790 and ST 612 804

This railway was opened by the Great Western Railway on 1 May 1903 from their main London – Bristol line at Wootton Bassett to junctions with the Bristol – Severn Tunnel line at Filton and Patch-way.

The new line shortened the distance between London and South Wales by 10 miles and relieved the tracks east of Bristol from the overloading which followed the opening of the Severn Tunnel – among other traffic great quantities of Welsh steam coal were being carried for the Navy at Portsmouth and the ocean liners at Southampton.

The crossing of the southern end of the Cotswolds entailed some substantial engineering works, the most notable of which are the Sodbury Tunnel, 2 miles 924 yd long (ST 793 812 to ST 752 815) and three massive brick viaducts at Winterbourne, of which the largest at ST 658 800 has eleven spans and is 265 yd long and 90 ft high.

---

20.   THE BRISTOL AND BATH   HEW 1061
TURNPIKE ROADS   ST 58, ST 98, ST 45 and ST 95
Illustrated on page 104

'Our shops, our horses' legs, our boots, our hearts have all benefited by the introduction of Macadam.' – Charles Dickens[37]

The system by which the construction and maintenance of roads was financed by the payment of tolls at 'turnpike' gates dates from the middle of the 17th century, but it was not until the beginning of the next century that the idea really took hold. Demands for speedier communications to serve government, the growing industries and the patrons of the newly popular spas and seaside resorts led to a great expansion which continued until the advent of railways in the 1830s.

Individual Acts of Parliament gave powers to Trustees to raise money by public subscription for the initial improvement of highways and for the installation of gates, tollhouses, weighbridges, milestones and signposts, and to charge users tolls for maintenance and the servicing of the debt. The trusts were served by a clerk, an accountant and one or more surveyors whose field of duty covered the area which could be ridden on horseback in a day.

The Acts for the Bristol (1727) and Bath (1707) Trusts were followed by amending Acts which eventually led to the Bristol Trust covering 178 miles of road, thus making it the largest trust outside London, and the Bath Trust, 73 miles.

Most of the major roads out of these cities today follow the routes of the turnpike roads, though inevitably, with improvements and the provision of by-passes over the years, there have been local deviations. Examples are the A38 (Gloucester and Bridgwater) and the A37 (Wells and Shepton Mallet) out of Bristol; the A4 (The Bath Road) towards London; and the A39 (Wells) and A367 (Shepton Mallet) out of Bath.

The Bristol and Bath Trusts are notable not only for having been among the earliest, but also for their association with the pioneering work of John Loudon McAdam, who was appointed General Surveyor to the Bristol Trust in 1816 and the Bath Trust in 1826, holding the former post until 1825 and the latter until his death in 1836.[38]

McAdam made his mark as much in the field of management as in that of construction.[39] He insisted on surveyors being responsible to the General Surveyor instead of, as hitherto, individually to the Trustees, and called for fortnightly reports on work done, with returns of labour, materials, transport and all costs, including contract work.

He was more a road-improver than a road-builder. On existing roads he had the top foot of stone taken up, where as much existed, broken into pieces not exceeding six ounces in weight and then relaid in two six-inch layers. For new roads he used the same grade of broken stone, in a single layer, sometimes as little as six inches thick. In this he differed from Telford, who insisted on a substantial foundation of coarse pitched stone, topped by a McAdam type surface layer.

McAdam's system was much cheaper than Telford's but the quality of his roads was nevertheless such as to permit, as in the case of Telford, a doubling of road speeds to as much as 12 miles per hour, which in turn was made possible only by a corresponding advance in the design and construction of coaches.

Horses' hooves and iron-tyred wheels ground the surface stones into dust which served as a binder, but this happy situation ended when motor vehicles came on the scene. Their rubber tyres sucked out the finer particles, broke up the road surface and produced clouds of dust. Dressing the surface with tar and sand and eventually the use of pre-coated stone was the remedy – hence 'tarmacadam'.

In 1888 the responsibility for public highways was transferred by Statute to Local Authorities. The turnpike system, with its gates and tollhouses, lapsed completely; but there are still many relics remaining. In 1967 the Bristol Industrial Archaeological

Society deposited with the City Museum a comprehensive index of tollhouses, milestones, boundary posts, finger posts, etc. resulting from an inspection of all roads within 40 miles of Bristol.

There is an interesting tollhouse, fronted by cast iron columns, at Ashton Gate, Bristol at the junction of North St and Ashton Rd (ST 572 718). This was originally at the beginning of the Bridgwater road but now, as the result of urban development, is virtually in a side street. At ST 597 636 on the B3130, which branches off the A37, a charming little two-storey tollhouse stands on an island at a road junction. At Red Post (ST 698 571) on the A367 at Peasedown St John is one of the Bath tollhouses, and diagonally across the crossroad there, a cast iron road marker dated 1827 is set at the foot of the wall of the Red Post Inn (1851).

---

21.   NEW BRIDGE, BATH   HEW 1107
ST 717 658

On the western outskirts of Bath an 85 ft span masonry arch carries the A4 trunk road over the River Avon.

First built around 1740, it was widened in 1828 as part of the turnpike improvements and at the same time the approaches were lengthened and flattened to ease the gradient over the bridge.

The new works incorporated a series of arches, largely for architectural effect, in the approaches at each end, so giving the south elevation, which is that seen from the approach road, an almost monumental effect. Less attention was given to the north elevation.

---

22.   VICTORIA SUSPENSION BRIDGE   HEW 811
BATH   ST 741 650
Illustrated on facing page

Apart from Clifton Suspension Bridge, the Victoria Bridge is the only survivor of eight suspension bridges of various designs, built across the River Avon in the Bath and Bristol area in the early decades of the 19th century. Designs, very much in the experimental stage, included the conventional chain of links with vertical suspenders (as for Clifton); stayed cantilevers, in which a series of straight ties splayed out from the towers to support the deck at intervals; and James Dredge's patent arrangement typified by the Victoria Bridge.

Victoria Suspension Bridge (R. A. Buchanan)

There are examples of Dredge's design, differing in detail, at Stowell Park, Wiltshire (HEW 378) and Aberchalder, Inverness-shire (HEW 888, NH 337 036).

The arrangement of the chains and suspenders is such that the tension in the chains is less in the middle of the span than at the towers. The varying slopes of the suspenders is an attractive feature.

Victoria Bridge, which was completed in December 1836, has a clear span of 139 ft 10 in. with a deck width of 17 ft 9 in. It was overhauled in the 1940s and is now used as a footbridge.

---

23.   GREEN PARK STATION, BATH   HEW 665
ST 745 648

This station has been disused for railway purposes since 1966 but the building has now been restored to its original condition to form part of a supermarket which a national grocery firm has developed on the site.

Opened on 1 May 1870 as the terminus of the Midland Railway branch from Mangotsfield on the Bristol–Gloucester line, and used jointly from 1874 as their Bath terminus by the Somerset and Dorset Railway, Green Park Station is outstanding among the lesser Victorian stations.

The Georgian frontage, with Ionic columns above the rusticated ground floor, a balustraded parapet, well-proportioned fenestration and a delicate iron porte-cochère in front of the entrance, is well in keeping with the architecture of Bath.

The train shed behind has fourteen bays of simple wrought iron girder arches of 66 ft 6 in. span, flanked by narrow aisles of segmental cast iron arches, all supported on slightly tapered octagonal cast iron columns. It is a good example of the pleasing effect of straightforward engineering built without extraneous decoration but to the right proportions.

An unusual feature of the station was the series of basements under the platforms and station buildings used as bonded stores and connected by underground passages to a Customs building which stood at the west end of the station yard.

| 24. | **BATH ROMAN WATER SUPPLY** | **HEW 1019** |
| | **AND DRAINAGE** | ST 752 648 |

Few visitors to the Roman Baths at Bath realize what lies beyond the steamy cavity from which emerges the hot water supplying the Great Bath.[40]

The original developers, in the 1st century AD, enclosed the hot springs in a roughly octagonal reservoir about 1600 sq.ft in extent, made by raising ashlar masonry walls to a height of about 8 ft. To get rid of water during construction openings were left, later to be plugged by 1 ft square oak blocks, some of which were visible when the reservoir was discovered in 1878.[41,42] The floor and walls were lead-lined.

A lead pipe 20 in. wide and 5 in. deep fed the Great Bath, which in turn fed two smaller baths.

Surplus water from the reservoir and the overflow from the baths were carried to the River Avon in a drain of rough masonry blocks, with a timber-lined channel at the bottom and furnished with manholes at intervals. Much of this sewer remains in use.

In the second and third centuries extensive alterations and extensions were carried out, involving more drainage and pipework. Some of the pipework can still be seen and much of the remainder can be traced by the recesses in which it lay in the paving round the baths.

Pulteney Weir

---

25. PULTENEY WEIR, BATH HEW 1272

Illustrated above ST 752 650

In mediaeval times a ferry crossed the Avon in Bath at the site now occupied by Pulteney Bridge (HEW 1060), the only bridge in Britain carrying buildings on both sides. Just below this a weir ran diagonally upstream from the right (west) bank and turned sharply in horseshoe form towards the left bank. At each end of the weir was a water mill.

The weir remained in much the same form after the mills had disappeared and after the building of the bridge in 1774.

Over the centuries Bath was subject to floods, aggravated by the presence down river of many weirs. In 1727 the Avon was made navigable as far up as Bath by means of six locks, for 74 ft by 16 ft barges, in the eleven miles from Hanham (ST 647 700), but this did nothing to improve the flooding situation.

In 1824 Thomas Telford was consulted and proposed channel improvements but surprisingly made no mention of the weirs. Nothing was done and this was the fate of several quite ambitious proposals over the next 130 years.

In 1965 the Bristol Avon River Authority undertook over

£2 million of improvement works through Bath from Pulteney Bridge to Saltford (ST 690 669) and about a tenth of the cost was for a new weir at Pulteney Bridge.[43] This is a sharply pointed horseshoe of three broad concrete steps springing from the right bank of the river and terminating downstream in a boat-shaped artificial island; between the artificial island and the left bank is the flood discharge channel controlled by a radial sluice gate.

The site of this imaginative scheme, in the midst of the architectural gems of Bath, is aesthetically sensitive. With the advice of Sir Hugh Casson the designers won a Civic Trust Award, Judicious landscaping has blended the sluice structure, very much in the modern idiom, into the environment, while the Award described the weir as a visual and acoustic triumph, with its three great stepped crescents of foam contrasting with the quiet waters above them reflecting Pulteney Bridge.

26. KENNET AND	HEW 1034
AVON CANAL	ST 754 644 to SU 470 672

The Kennet and Avon Canal, built between 1794 and 1816 with John Rennie as Engineer, runs 57 miles from the Avon at Bath to the Kennet Navigation at Newbury, via Devizes and Hungerford, and afforded a more usable route between London and the Severn than the Stroudwater (HEW 292) and the Thames and Severn (HEW 1139) Canals. (See Chapter 5.)

The opening of the Great Western Railway in 1841 rapidly affected its fortunes and by the early part of the 20th century it had largely fallen into disuse. In recent years a lot of restoration work has been carried out. Much of the original construction can be seen.

There are more than 70 locks; five masonry aqueducts, four in the classical style; a 502 yd tunnel at SU 236 632 (HEW 1077) and two very short ones in Sydney Gardens, Bath; and numerous bridges in timber, cast iron, brick and masonry, notable among the last being the decorated Ladies Bridge at Wilcot (HEW 895, SU 135 610). The suspension bridge in Stowell Park (HEW 378, SU 146 614) is by Dredge, who also built Victoria Bridge, Bath (HEW 811). The nine timber swing bridges were rebuilt in the 1960s to the original design, with ball bearing pivots – as at Rusty Lane (HEW 385, SU 146 614). There are two pumping stations, at Claverton (HEW 1076) and Crofton (HEW 57).

Near the spot where the canal leaves the Avon was the site of

Dolemead Wharf, to which stone was brought down from Combe Down quarries, $1\frac{1}{2}$ miles to the south-east of Bath, by Ralph Allen's Waggonway (HEW 1338, ST 752 643 to ST 758 622), built in 1730 and probably the first railroad in Britain to have been described in print, it having been the subject of an article by Charles de Labelye in Desargulier's *Experimental Philosophy*.[44]

27.	WIDCOMBE LOCKS,	HEW 1110
	BATH	ST 754 643 to ST 758 646

Widcombe Locks lift the Kennet and Avon Canal through 66 ft 6 in. from its junction with the River Avon at Bath to the side slopes above the river. Originally seven, there are now six locks, each about 75 ft long and just over 14 ft wide, built to take 50-ton barges, 70 ft long and 13 ft wide. Each lock is furnished with a side pound dug out of the hillside.

They were begun in Bath stone against Rennie's advice, but were completed in a more durable stone from Winsley quarry. The locks have been well maintained with repairs done in red or blue engineering bricks. In 1971 a major restoration was carried out.

The hand-operated timber gates have built-in sluices, those in the lowest gates having been made hydraulic-assisted during the restoration.

A charming feature of the locks is the pair of tiny cast iron footbridges, over the Top Lock and Wash House Lock, built by Stothert of Bath circa 1815 (HEW 360, ST 758 646 and 806, ST 757 644).

A similar bridge, of 27 ft span (HEW 649) carries a footpath over Kings Weston Lane in Bristol (ST 545 773).

28.	CAST IRON BRIDGES,	HEW 808, ST 758 653
	SYDNEY GARDENS,	HEW 807, ST 758 653
	BATH	HEW 539, ST 758 652

A short way northward from Widcombe Locks may be seen three more cast iron arched bridges.

The first (HEW 808) is a footbridge of about 23 ft span and 10 ft wide, having four cast iron ribs in two sections with cast iron deck plates. The ribs have ring pattern spandrel filling. The fencing is of simple palings with three rails, alternate palings being full height and three-quarter height. This bridge was overhauled in 1978.

The second (HEW 807) is very different. It spans just over 30 ft on a pronounced skew and carries a 19 ft wide roadway. The face ribs are solid panelled, in two sections, and there are five interior ribs of T or cruciform section carrying solid cast iron deck plates 10 ft wide set parallel to the canal. The parapets have distinctive cast iron panels of a large diamond superimposed on a cross.

The ironwork for both bridges, which date from 1800, was cast at Ironbridge.

Access to them is from Sydney Road alongside a building erected above Sydney Gardens No. 1 Tunnel (and once housing canal offices) and back under the building through the tunnel.

The third bridge (HEW 359) is much later, built in about 1865 over the Great Western Railway (HEW 1017). The span of 30 ft 8 in. reflects the width of railway formation required to accommodate Brunel's broad gauge tracks. Some 50 yd away towards Bath a skew masonry bridge carrying a road over the railway bears a plaque to 'Isambard Kingdom Brunel 1841'.

---

29.   CLAVERTON PUMPING STATION   HEW 1076
Illustrated below   ST 791 644

Claverton water-powered pumping station started work in March 1813 lifting water from the River Avon to the Kennet and Avon Canal 53 ft above.

Claverton Pumping Station

Originally a breast wheel, 25 ft wide, 18 ft 4 in. dia., fed through a leat from the river, operated the two bucket pumps through two gear wheels to a common crankshaft from which in turn cruciform section connecting rods work the pumps. The larger gear wheel has wooden teeth.

In 1858 the Great Western Railway, who then owned the canal, replaced the single wheel by a pair, each 15 ft 6 in. dia., 11 ft wide, but still on one shaft, and introduced an intermediate bearing on a cast iron A-frame.

In 1952 some of the wooden teeth stripped and a small diesel pump was installed to maintain the water supply.

The pumps are housed in a masonry building, with the water wheels under a timber structure with a tiled roof.

Restoration of the machinery has been carried out by the voluntary efforts of Bath University and the Kennet and Avon Canal Trust.

In 1981 the British Waterways Board installed two 75 hp electric pumps just upstream of the station and presented the diesel pump to the Trust, for preservation.

---

30.   DUNDAS AQUEDUCT   HEW 288
Illustrated on page 116   ST 785 625

The Dundas aqueduct lies in the Limpley Stoke Valley, south-east of Bath, and carries the Kennet and Avon Canal over the River Avon. There are right-angled bends in the canal at each end and a turning basin, which also served as the terminal basin of the now long defunct Somerset Coal Canal, was provided at the Bath or west end.

The main semi-circular 65 ft span stone arch over the river is flanked by two tall elliptical arches of 19 ft 3 in. span.

The aqueduct was begun in 1796, completed in 1798 and opened in 1800.

No-one would tender for the foundations, which were built in cofferdams by direct labour. James McIlquham was the contractor for the superstructure.

With its exterior cornices on both sides, and stone balustrade, the aqueduct forms a monumental piece of architecture. It is a fine example of the work of John Rennie, who surveyed the route of the canal from Bath to Newbury in 1788, at the age of 29 and was appointed Engineer for the canal in 1794.

The aqueduct is named after Charles Dundas, Chairman of the

Canal Company from its inception until his death, as Lord
Amesbury, in 1832.

31. DEVIZES LOCKS   HEW 1078
Illustrated on facing page   ST 966 617 to SU 000 617

West of Devizes, 29 locks raise the canal from Lock 22, west of the
B3101 near Rowde to Devizes Top Lock 50 at Town Bridge. They
are to be compared with the locks on the Worcester and Birmingham Canal at Tardebigge (HEW 768, Chapter 5) but although
these have the greatest number of locks in one flight in Britain, the
Devizes locks rise through a greater height (237 ft) over a lesser
distance (about 2 miles).

Locks 28 to 44 form the remarkable Caen Hill flight, where 17
locks, on a gradient of 1 in 30, are separated from one another only
by the entrances to the closely spaced side pounds on the north side
of the canal.

In 1829, some years after the opening of the locks, gas lighting
was installed and an extra charge was made for craft passing
through at night with the use of the lighting.

The locks were closed in 1951, but restoration is in hand as a

Dundas Aqueduct

joint endeavour by Kennet District Council, the British Waterways Board and the Kennet and Avon Canal Trust, and is making good progress.

On the west side of Prison Bridge, Devizes, is a tablet commemorating John Blackwell, one of Rennie's resident engineers, who went on to serve the canal company as Superintending Engineer for 34 years until his death in 1840.

32. CROFTON PUMPING STATION   HEW 57
Illustrated on page 118   SU 262 622

Situated about two miles west of Great Bedwyn, this station was built to feed water to the summit level of the canal via a leat from Wilton Water, some 300 yd south, to a pound west of Crofton Top Lock. In contrast with Claverton Pumphouse (HEW 1076), it was steam-powered.

The building houses two 19th century engines. The 1812 engine, 42 in. bore, 8 ft stroke, is the only Boulton & Watt beam engine still under steam and doing the work for which it was installed and lays claim to be the oldest working beam engine in the world. The second engine, a Sims combined cylinder engine,

Devizes Locks (British Rail)

40 in. and 22 in. bore and 8 ft stroke, built by Harveys of Hayle, was installed in 1845 to replace an even earlier (1809) Boulton & Watt engine.

Both engines have been restored by the Kennet and Avon Canal Trust and operate on several occasions each year when the station is open to the public.

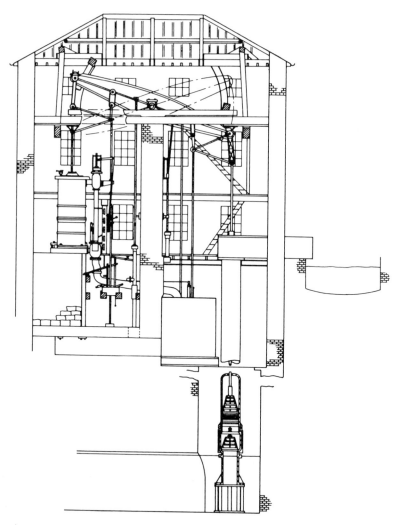

Crofton Pumping Station: Boulton & Watt 1812 engine (Kennet and Avon Canal Trust)

---

33. GREAT BEDWYN SKEW BRIDGE HEW 379
SU 276 637

At Great Bedwyn the Kennet and Avon Canal is crossed by an arch of about 25 ft 3 in. span carrying an unclassified road.

Said to be one of the first skew arches ever to be built in brick, it has certain apparent irregularities in the coursing of the brickwork which have led to the charge that John Rennie, the designer, had not fully grasped the principles.

However, like the non-skew bridges along the canal, the width is greater at the springing than at the crown. This feature would have introduced problems in the layout of the bricks as compared with the case of a parallel-ended barrel.

---

34. SILBURY HILL HEW 510
Illustrated below SU 100 685

Adjoining the north side of the A4, some six miles west of Marlborough, Wiltshire, and close to the Avebury Stone Circle (SU 101 700) is the largest prehistoric man-made mound in Europe. Dating from 2660 BC it is roughly 430 ft dia., 130 ft high, with side slopes of 1 in $1\frac{1}{2}$, and is surrounded by a 30 ft wide ditch. Exploratory tunnelling over the past two hundred years has established that it is not a burial mound as it was once thought to

Silbury Hill (B. T. Batsford Ltd)

be, but although its purpose remains unknown it has been shown that the construction was quite sophisticated for its time.[45]

Over and around an original 100 ft dia. mound was built a series of radial and concentric walls of chalk blocks which were filled in with chalk rubble to form a number of 17 ft high steps and benchings. These were then packed with rubble and the whole mound turfed over.

The borrow ditch left the base with a 30 ft high vertical face of virgin chalk, against which was piled chalk rubble in 2 ft deep stepped layers with timber revetment, to protect the face against weathering and slipping. This protection is now buried in the silt which has partly filled the ditch over the centuries.

Even by modern standards it was an impressive undertaking. It has been estimated that five hundred people, with the most primitive digging tools and carrying materials in baskets, would have taken ten years to complete the work and that in relation to the likely available population the enterprise was, for its time, comparable to the American or Russian space programmes.

CHAPTER 5

Gloucestershire and Hereford and Worcester

This region was not greatly affected by the Industrial Revolution which from the 1760s developed rapidly in the West Midlands just to the north. Consequently main towns such as Hereford, Evesham and even Gloucester still perform their ancient functions as commercial centres for a predominantly rural area.

However, timber, iron ore and coal (mined until recent years) in the Forest of Dean supported some local metal working industry as well as being exported to other parts of the country through the small port of Lydney.

The first brass making in Britain was set up in the Forest in the 16th century before being transferred to Bristol. South of Gloucester a number of textile mills survive out of the once extensive woollen industry based on Cotswold sheep.

Communications from earliest times were dominated by the Severn running from north to south through the area. Development of the industrial Midlands required links from the river to the higher ground to the north-east. The Staffordshire and Worcestershire Canal (1772) and the Worcester and Birmingham Canal (1815) are typical of these. The Stroudwater and Thames and Severn Canals aimed at joining the Severn and the Thames but were largely superseded by the Kennet and Avon Canal further south.

The Gloucester and Sharpness Canal, by cutting off a dangerous loop of the Severn, brought about rapid expansion of Gloucester as a port, followed late in the 19th century by Sharpness Docks.

After the canals came the railways, with trunk routes running north and south and east-west lines leading into Wales.

In more recent times the M50 and M5 motorways have introduced a new element to the landscape.

It is not surprising therefore that many of the interesting civil engineering works in this area are related to transport by one means or another.

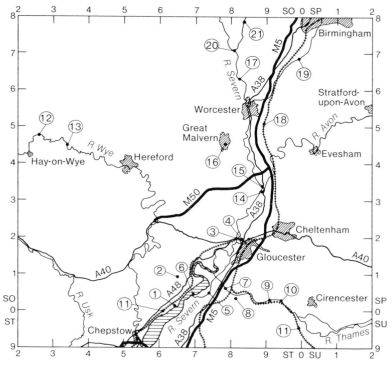

1. Lydney Docks
2. Roman Road, Blackpool Bridge, Forest of Dean
3. Over Bridge, Gloucester
4. Gloucester Docks
5. Gloucester and Sharpness Canal
6. Severn Railway Bridge
7. Stroudwater Canal
8. King's Stanley Mill, Stonehouse
9. Thames and Severn Canal
10. Sapperton Tunnel, Thames and Severn Canal
11. The railway from Swindon to Severn Tunnel Junction via Gloucester
12. Whitney Toll Bridge
13. Bredwardine Bridge
14. Quay Pit Bridge, Tewkesbury
15. Mythe Bridge, Tewkesbury
16. Great Malvern Station
17. Holt Fleet Bridge
18. Birmingham and Gloucester Railway
19. Tardebigge Locks
20. Stourport-on-Severn Canal Terminus
21. Stour Valley Hydraulic Arch

1. LYDNEY DOCKS

HEW 626
SO 634 018 to SO 651 014

Lydney was a port from the reign of Henry II but over the centuries siltation of its small river gradually reduced its usefulness.

In 1810 the Severn and Wye Tramroad Company, who operated 30 miles of tramroad in the Forest of Dean, built an entrance lock from the Severn and a basin 760 ft long and 105 ft wide, with a canal 3050 ft long leading to an upper basin 1100 ft long and 90 ft wide to take 100 ft by 24 ft barges. In 1821 a tidal basin and outer lock were added, which enabled ships up to 400 tons to use the lower basin. So Lydney became the principal sea outlet for coal mined in the Forest, and at their peak the docks handled as much as 400 000 tons in a year.

Ownership passed to the Great Western Railway and the Midland Railway jointly in 1894, and after nationalization of the railways in 1948 to the British Transport Commission and eventually to the British Transport Docks Board. With the cessation of mining in the forest the coal trade disappeared, and of the banks of coal sidings which ran along the docks to serve nine coaling stages there is now little trace.

Since August 1980 the docks have been owned by the Severn-Trent Water Authority who intend to develop the area for small craft and as a public amenity.

---

2. ROMAN ROAD, BLACKPOOL	HEW 630
BRIDGE, FOREST OF DEAN	**SO 653 087**

This fragment of the Roman road between Lydney and Mitcheldean (Ariconium) was built in the first or second centuries AD and lies alongside an unclassified road off the B4431 about two miles from Blakeney and leading northward to Upper Soudley.

The remains extend for about 85 ft and include part of a road junction. Between two lines of edging stones 4 in. to 6 in. wide and 12 in. to 18 in. long is random paving of rough stones up to about 18 in. by 12 in. in plan. The width, just under 8 ft, suggests that the road was an Actus, for a single carriageway. This was the second class of Roman road, the first class being the Via, a two-lane road 14 ft or more wide.

The Romans built their roads with military considerations in mind, in straight lines from one natural obstacle to another, a strip on each side being cleared of trees and bushes to guard against marauders. Construction was in layers, usually on an embankment, the 'agger', dug from borrow-ditches or pits alongside. The material used depended on what was locally available. The thickness could vary from three inches up to two feet according to the weight of traffic expected. On the most heavily used roads there would be a layer of small stones or gravel, followed possibly by stone – lime concrete or, in iron-working districts, iron slag, worked into a firm mass. Finally, the wearing surface, laid between edging stones, consisted of roughly rectangular or polygonal stones, several inches thick, closely fitted together. The road surface was therefore raised above the surrounding ground and with the heavy camber provided, as much as a foot in a width of 14 ft, formed a well-drained highway.

---

3.   OVER BRIDGE, GLOUCESTER
Illustrated below

HEW 148
SO 817 196

Immediately to the south of the present bridge which carries the A40 trunk road across the Severn about one mile west of Gloucester, a masonry arch of 150 ft span crosses the river.[1] It was completed in 1828. Thomas Telford was the Engineer,[2] and he based the detailed design on Perronet's five-span bridge across the Seine at Neuilly (1774, replaced 1956).

The main soffit is built to an elliptical curve with 35 ft rise, but the voussoirs on the face follow a segmental curve of only 13 ft rise and spring 22 ft above the main springing. This produces a funnelled entry upstream and downstream, aimed at easing the passage of flood water through the bridge, and at the same time gives an interesting complex shape to the arch.[3]

The abutments rise from timber platforms lying on gravel 27 to 33 ft below ground. The wing walls are founded 8 to 10 ft deep, directly onto the ground.

When the arch centring was struck, the crown sank 2 in., and later movement of the wing walls produced a further sinking of 8 in. Telford blamed this on his parsimony in omitting piling or platforms under the walls, and considered the work as being one of his failures.

Over Bridge, Gloucester (Telford: *Life and atlas*)

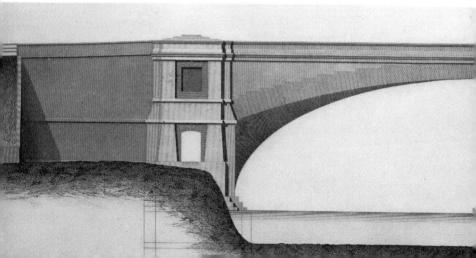

Nevertheless, the bridge, which had a 17 ft roadway and two 4 ft footpaths, carried traffic on the A40 until it was replaced by the new bridge just upstream in 1975.

It now stands in isolation, as a scheduled Ancient Monument.

The A40 London to Fishguard road (HEW 1198) between Gloucester and Carmarthen features in Chapter 3. In the other direction, just east of Cheltenham, it passes alongside Dowdeswell Reservoir (HEW 1306, SO 99 19) built in 1878 by J. F. La Trobe Bateman and similar in layout to his greatest work, the Longdendale scheme in the Etherow Valley near Manchester.[4]

4. **GLOUCESTER DOCKS**	**HEW 625**
Illustrated on page 127	SO 827 183

Gloucester has been a river port since Roman times. Its 12th/13th century developments are remembered by the street known as The Quay.[5,6]

South of this the present docks comprise the Main Basin (1812), a Barge Arm (1824), two dry docks (1834 and 1851), the Victoria Dock (1849) on the east side of the basin and Monk Meadow Dock (1890) on the west side of the Gloucester and Sharpness Canal (HEW 466), which enters the south end of the basin and which is widened over its last $1\frac{1}{2}$ miles to afford berthing facilities. At first, access to the docks was solely through a lock from the Severn at the north-west corner of the Main Basin, but with the opening of the canal in 1827 the use of the lock became confined to barge traffic to and from the upper reaches of the river.

The docks are notable for nearly a score of multi-storey warehouses, the majority built in the first half of the 19th century

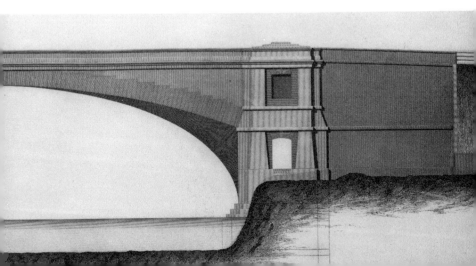

to meet the rapid increase in the trade in imported corn which followed the relaxation and subsequent repeal of the Corn Laws.[7] The earliest dates from 1827. All follow the same general pattern: brick walls and slate roofs, with timber floors carried on rows of cast iron tubular columns, loading doors on all floors and hand-operated hoists projecting from the roofs. The Pillar Warehouse (1849), a Grade II listed building on the east side of the canal just south of the docks is of particular interest, with the three upper floors on the waterside hanging over the quay and supported on a colonnade of seven cast iron Doric columns.

Although traffic declined steadily during this century, the docks continue to trade and the owners, British Waterways, are co-operating with the City Council to preserve the docks not only as a commercial feature of the city but also as a centre for leisure and cultural activities.

---

5. GLOUCESTER AND SHARPNESS CANAL
Illustrated on page 128

HEW 466
SO 827 185 to SO 668 031

This, the first ship canal in Great Britain, was built between 1798 and 1827 to the design of Robert Mylne, with Thomas Telford as adviser to the Canal Commissioners. The Pinkerton family, who worked on canals and navigations from Yorkshire to South Wales, were involved as contractors. Eighteen feet deep and 86 ft 6 in. wide, it was meant for 400 ton ships, but modern ship design permits the passage of vessels up to 1200 tons, with a maximum beam of 29 ft which is determined by the bridge openings.

The canal connects with the Severn through a lock at Gloucester (HEW 625) and also at Sharpness where the sea lock leads to a basin from which two locks, the smaller of which is now partly filled in, lead to the canal. The basin and lock structures are basically as originally built and there is an attractive Lockmaster's House.

In 1874 a new dock at Sharpness was opened (HEW 388, SO 673 024). Entrance gates 57 ft wide lead to the Tidal Basin from which a 320 ft by 57 ft lock, with 22 ft of water on the cill, connects to the 20 acre dock, some 2000 ft long which can accommodate 7000 ton ships. At the inner end of the dock a junction was made with the canal, and canal traffic now uses the dock entrance. The canal from this junction to the original basin is used as a marina.

At Saul Junction (SO 756 093) the crossing of the Stroudwater Canal (HEW 260) is of interest, and is substantially as it was in

1820. There is a small dry dock, built in 1827, which is still used by a firm engaged in the repair and building of small craft.

All the swing bridges, of which there are fifteen, the lockgates and equipment are of recent date but several of the original elegant Adam style bridgekeepers' cottages still remain.

6. SEVERN RAILWAY BRIDGE   HEW 338
SO 679 033 to SO 670 041

Just above Sharpness stand the masonry pivot tower and east abutment of the swing bridge built in 1874–76 to carry a railway across the canal as part of the Severn Railway Bridge.[8–10]

This was one of the longest railway bridges in the United Kingdom and carried a single track to link the Midland Railway Birmingham–Gloucester–Bristol line with the Great Western Railway Gloucester–Chepstow–South Wales line.

There were 21 spans of wrought iron bowstrung trusses, ranging from about 125 ft to 325 ft, on cast iron cylinder piers, together with a 12 arch masonry viaduct at the west end and the swing span over the canal at the east end.

Gloucester Docks (Russell Adams, Gloucester)

Two spans of the bridge were brought down in 1960 by a petrol barge colliding with one of the piers in fog, and eventually the whole structure was demolished in 1968 – 69.

7.   STROUDWATER CANAL   HEW 292
Illustrated below   SO 751 104 to SO 848 050

This canal is of interest because the company formed in 1730 to promote it can lay claims to being the oldest canal company still in being, as although the canal was abandoned in 1954 a small revenue continues to be drawn from fishing rights.

Opposition from mill owners on the River Stroudwater (or Frome) delayed the start until 1775, although between 1759 and 1763 there was an attempt, which proved too costly in operation, to form a Navigation of the lower reaches of the river and to bypass the mills by the novel device of making the new channel with dead ends overlapping the millponds and transferring goods from one section to the next by crane.

The 8¼ mile canal, completed in 1779 to plans by Thomas Yeoman, ran parallel to the river from Framilode on the Severn to

Saul Junction: Gloucester and Sharpness Canal and Stroudwater Canal

Wallsbridge in Stroud whence it was later continued by means of the Thames and Severn Canal (HEW 1139). It was 42 ft wide and 6 ft deep, and twelve locks dealt with the total rise of 102 ft.[11]

A good deal of the waterway remains, but few structures, of which the most interesting is the lock at Saul Junction built in 1820 to take the canal across the Gloucester and Sharpness Canal (HEW 466).

The Stroudwater and Thames and Severn Canal Trust are doing some restoration work but there is no possibility of the canal being restored throughout as a section near the middle was obliterated by major roadworks on the M5 motorway and its junction with the A38.

---

8. KING'S STANLEY MILL, STONEHOUSE

HEW 426
SO 813 043

Illustrated on page 130

Near the Stroudwater Canal, this five-storey textile mill, built in 1813 on a site with continuous records of mills going back to 1563, with a mention in the Domesday Book, is still working and producing fine woollen cloth and industrial fabrics.[12]

The greater part of the mill was built to be completely fireproof and incorporated no combustible materials, and even had iron doors.

It differs from other mills in the area in having the walls above first floor of brickwork, instead of being in Cotswold stone throughout. The elevations include a number of handsome 'Palladian' cast iron windows.

Internally the floors are carried on arcades of cast iron columns which support an elaborate system of ornate cast iron arches. These in turn carry cast iron floor beams between which are brick jack arches. Over these is a filling of ash or rubble on which the stone floor slabs are laid.

The ironwork was cast at Dudley, Staffordshire, and brought to site along the canal whence it was hauled for a distance of about $\frac{1}{3}$ mile over the fields. In view of their size and number the accuracy of the castings is remarkable, in that the seatings for the bearings for the complex system of line shafting are integral with the castings, with limited scope for adjusting the alignment.

During the 19th century steam engines were installed to augment the original six water wheels, which in 1867 were replaced by a water turbine, scrapped in 1968. The mill is now wholly electric.

9.   THAMES AND SEVERN   **HEW 1139**
CANAL   SO 848 050 to SU 205 988

Completed in 1789 to the design of Robert Whitworth, with Josiah Clowes as engineer in charge, this canal joined the east end of the Stroudwater Canal (HEW 292) to the Thames just above Lechlade and completed the first waterway link between Thames and Severn.[13] There were 44 locks in its $28\frac{3}{4}$ mile length. At its west end Severn trows could reach Brimscombe where there were quays and

King's Stanley Mill, Stonehouse (Eric de Maré)

warehouses and facilities for weighing boats. The rest was for narrow boats.

At Thames Head (SU 988 988) the canal was fed from the springs at the source of the Thames, pumped up first by wind power, then by a Watt steam engine installed in 1802 and replaced in 1854 by a Cornish engine.

The canal was abandoned in 1927–33 and much has completely disappeared, but there are still some remains to be seen, mainly locks and overbridges, and five interesting canal keepers' cottages.

These were circular three-storey stone buildings; a store and a stable on the ground floor; a 16 ft 10 in. dia. living room reached by outside steps; stairs between inner and outer walls leading to the bedroom; and a roof in the form of an inverted cone designed to catch rainwater. Two are still lived in, at Cerney Wick (SU 079 960) and Inglesham Lock on the Thames (SU 205 988).

---

10. SAPPERTON TUNNEL, THAMES   HEW 629
AND SEVERN CANAL   SO 943 033 to SO 966 006

Illustrated on page 133

The Sapperton Tunnel lies under the A419 Stroud–Cirencester road, between Sapperton and Coates in Gloucestershire.

Built between 1783 and 1789 in the summit level of the canal, it was at the time the longest tunnel (2 miles 297 yd) in the country.

Both portals, easily accessible, lie in pleasant wooded valleys. The west portal is of plain limestone masonry. The east portal, recently restored by the Stroudwater and Thames and Severn Canal Trust, is a striking structure in Cotswold stone in the classical style. The work was designed by Robert Whitworth and was built by five separate gangs of workmen.

There was no towpath and boats were propelled by the crew lying on their backs and pushing against the sides and roof with their feet.

Access through the tunnel is no longer possible due to roof falls.

---

11. RAILWAY FROM SWINDON   HEW 1179
TO SEVERN TUNNEL   SU 146 851 to ST 460 875
JUNCTION VIA GLOUCESTER

The area covered by this chapter is traversed by much of the route of this railway line which began as the Cheltenham and Great

Western Union Railway, a broad gauge line authorized in 1836, engineered by I. K. Brunel, and built under the direction of his chief assistant, R. P. Brereton.

The line diverges northward from the London–Bristol railway (HEW 1071, Chapter 4) a short way west of Swindon station. At SO 960 013 it crosses the line of the Sapperton canal tunnel (HEW 629) on the Thames–Severn Canal (HEW 1139). Soon after, come the Sapperton railway tunnels (HEW 1211, SO 940 022). These were envisaged as one curved 2800 yd long tunnel at low level, but after nine shafts had been sunk and the pilot heading nearly completed, in 1841, financial stringencies obliged Brunel to take the drastic action of raising the track some 90 ft, which resulted in two tunnels separated by a very short cutting next to the A419 road.

Next comes the Frampton Mansell Viaduct (HEW 389, SO 919 028), and the line then follows the Golden Valley down to Stroud, passing through the Chalford Gorge, where railway, road, river and canal run close together. After Stroud a sharpish curve northward out of the valley through Stonehouse leads to the junction made on 12 May 1845 with the Bristol and Gloucester line (HEW 1075) at Standish (SO 803 069). About $1\frac{1}{2}$ miles beyond the junction the line is crossed by two interesting cast iron bridges (HEW 935, SO 810 090).

The Bristol & Gloucester Railway[14] began as the Bristol and Gloucestershire Railway, six miles of standard gauge horse-drawn tramroad from the River Avon, opened in 1835 to serve the North Bristol coalfield.

Under an Act of 1839 an extension, with steam traction, was authorized to Standish Junction, and the name was changed to the Bristol and Gloucester Railway with broad gauge tracks and running powers over the line from Standish to Gloucester. At Gloucester the broad gauge lines met the standard gauge Birmingham and Gloucester (HEW 1091). The need to transfer goods, as well as passengers, at the break of gauge soon demonstrated in a practical fashion what had hitherto been a matter for theoretical discussion, namely the disadvantage of having more than one gauge in the national system.

In 1845 the Midland Railway acquired both the Birmingham and Gloucester and Bristol and Gloucester lines and by 1854 had added standard gauge throughout the latter.

An interesting sidelight on the battle of the gauges is that in 1832 the Kennet and Avon Canal Company opened the Avon & Gloucestershire Railway (HEW 1074), a standard gauge horse tramway from near Keynsham (ST 666 693) to a junction with the Bristol & Gloucestershire at Mangotsfield (ST 672 756) with running powers over a further 2½ miles northward. Engineers of the new Bristol & Gloucester had therefore from the outset to lay standard gauge over this length to accommodate the Avon and Gloucestershire and so established the first example of mixed-gauge track in Britain.

Over the route so far described the 'Cheltenham Flyer' was in the 1930s making the headlines with daily fast runs to London and back, albeit at speeds much less than are nowadays taken for granted on the High Speed Trains in various parts of the country.

The line to Gloucester leaves the Gloucester–Birmingham line at Tuffley (SO 845 175). After Gloucester station it passes close to the Cathedral, prominent on the left, and crosses in quick succession five roads and the two arms of the Severn, the second of these at SO 816 195 just downstream of Telford's Over Bridge (HEW 148). The first 7½ miles after Gloucester, to Grange Court (SO 725 161), were Great Western from the outset; beyond that the line is that of the South Wales Railway (HEW 1199).

To Chepstow the right bank of the Severn is followed, past the site of the Severn Railway Bridge (HEW 338) and across Lydney Harbour (HEW 626). After Chepstow the line passes under the M4 soon after the latter comes off the Wye and Severn Bridges

Sapperton Canal Tunnel: east portal

Whitney Toll Bridge

(HEW 201), passes Portskewett, at the western end of the Bristol and South Wales Union Railway (HEW 1033), and the Sudbrook Pumping Station of the Severn Tunnel (HEW 232), and near the tunnel entrance crosses the Main Line from London and Bristol, which joins it at Severn Tunnel Junction. (See Chapter 4.)

The South Wales Railway opened from Chepstow to Swansea in 1850 and from Grange Court to Chepstow East in 1851, but the final link across the Wye at Chepstow, Brunel's iron bridge, was not completed until 1852. The design of the 300 ft main span, with suspension chains tied by tubular arched girders, was not unlike that of his famous Saltash Bridge (HEW 29). The plate girder approach spans were replaced in 1948 and in 1962 the successors of the firm which built the original bridge replaced the main span by welded underline Warren girders.[15]

Although the branch lines such as those to Tetbury and Cirencester (where the station building is incorporated in the town's bus station) and the several feeder lines from the Forest of Dean collieries have long since closed, the route remains well in use as part of the national rail network.

---

12. WHITNEY TOLL BRIDGE   HEW 816
Illustrated above   SO 259 474

Whitney Toll Bridge Act of 1797 authorized the building of a bridge over the River Wye at Whitney, near Hay-on-Wye. The bridge was washed away twice, and the present structure dates from about 1820.

It is unusual, consisting of a stone arch at each end with a centre section 110 ft long built of timber. The timber section is in three roughly equal spans, supported on two timber piers, themselves founded on masonry sub-piers. There are diagonal timbers running from the piers to the bridge beams to give additional support. The roadway is constructed of macadam surfacing laid on a timber deck and is 14 ft wide. The bridge currently carries a ten ton weight restriction.

A small toll house is situated at the north end of the bridge and is in use. The toll rights are privately owned, although the bridge serves the junction of a public highway (B4350) with the A438.

---

13. BREDWARDINE BRIDGE

HEW 866
SO 336 447

Bredwardine Bridge, considered to be one of the finest brick bridges in England, carries a minor road over the River Wye near the Herefordshire village of Bredwardine, and has stood almost unchanged since it was built in 1769 to replace two ferries.

Six semicircular arches, of about 32 ft span, supported on brick piers, carry a 13 ft wide roadway, the total length of the bridge being 285 ft. The piers have triangular cutwaters both upstream and downstream, and on two of the piers they are continued upwards to form pedestrian refuges in the parapets. Only the four middle arches are over the river; the other two serve as flood-relief spans.

It was originally a toll bridge. Tolls were removed in 1824 but reimposed in 1863 to help the Trustees with the cost of repairs. In 1894 Hereford County Council assumed responsibility for the bridge and in 1922 carried out major repairs to the foundations and arches.

---

14. QUAY PIT BRIDGE, TEWKESBURY

HEW 387
SO 892 330

At the end of a short side turning off the A38 in the middle of Tewkesbury, a 52 ft span bridge crosses the River Avon just above the locks which join that river to the Severn.

Five cast iron arched ribs with a rise of about 11 ft spring from water level, their spandrels infilled with circular hoops.

The bridge is decked with cast iron plates and the outer ribs are surmounted by solid cast iron kerbs which carry cast iron railing fences, with ornamental features at the middle. The bridge is thought to have been built circa 1820, but may be earlier.

In recent years the deck level has been raised by adding metalling and tarmacadam, while additional kerbs, limiting the road width to 19 ft 6 in., keep vehicles from contact with the fencing. The bridge is used by heavy lorries to and from the flour mill across the river.

---

15.   MYTHE BRIDGE, TEWKESBURY   HEW 134
Illustrated on facing page   SO 889 337

This structure[16] is described in an original communication of 11 March 1828 to the Institution of Civil Engineers by the Engineer, Thomas Telford himself, who considered this bridge to be 'rather special'.[17]

It is one of several similar bridges cast by William Hazledine of Shrewsbury and has a main span of 170 ft, consisting of six cast iron arched ribs, each of eight segments of five panels of open web X type. The spandrels also have open X pattern bracing with cross bolts and distance pieces at the intersections. The deck is of cast iron flanged plates: the ballast plates are solid panels with the date 1825 cast on the centre ones, and there is a vertical paled parapet with stone pilasters. The width is 24 ft including two narrow footpaths.

The very substantial abutments are piled and incorporate six tall Gothic pointed arches of 12 ft span each side.

The bridge carries the A438 Ledbury road over the River Severn just west of Tewkesbury, on the county boundary, and has a load restriction despite being strengthened with a reinforced concrete deck slab in 1923.

---

16.   GREAT MALVERN STATION   HEW 1113
                                                          SO 784 457

This station, opened in 1861 on the line between Worcester and Hereford, was provided by its architect, E. W. Elmslie, with a magnificent cast iron platform roof. Cast iron columns support decorated beams and cantilevers running parallel and perpendicular to the platform edge. There is a pitched glass roof and a timber canopy.

Of the 28 columns (14 on each platform), 26 are intricately decorated at the column heads by ornate castings representing the leaves and fruit of trees. There is a large variety of different patterns and all are brightly painted.

Railway stations are on the face of it purely architectural works. They have, however, a wider function. They are the point of contact between the public and the railway, in which respect they compare with seaports and airports. Moreover the platforms, track layouts and often the ancillary buildings are essentially engineering works.

17. HOLT FLEET BRIDGE   HEW 135
SO 824 634

Holt Fleet Bridge, built in 1828, carries the A4133 road across the River Severn in a single span of 150 ft and was designed by

Mythe Bridge (Ironbridge Gorge Museum Trust)

Thomas Telford. The arch is made up of five cast iron ribs each 3 ft 2 in. deep made in seven segments. The deck is supported from the arch by cast iron struts inclined to the vertical and arranged in intersecting pairs. The struts are cross-connected at the point of intersection by tie rods with distance pieces. They are secured by mortise and tenon joints with cast iron wedges. At deck level, cast iron beams 6 in. deep and 2 in. wide carried the deck plates and the roadway. The massive sandstone abutments of the bridge are pierced on each side of the river by flood arches.

In 1928 the bridge was strengthened by encasing the upper and lower edges of the arch ribs in reinforced concrete extending across the whole width of the bridge. In addition, one of each pair of struts was encased in concrete. A new reinforced deck slab was cast on new cross beams and the roadway was widened by cantilevering out over the sides of the old arch.[18]

---

18. BIRMINGHAM AND   HEW 1091
GLOUCESTER RAILWAY   SP 070 867 to SO 838 185

In 1832, Brunel made a survey for a broad gauge railway to link Birmingham and Gloucester but his route was not chosen, an alternative standard gauge route proposed the following year by Captain W. S. Moorsom being preferred by the promoters. The Birmingham and Gloucester company was incorporated on 22 April 1836. The railway was built from Birmingham to Cheltenham, where it made use of the line of the existing Gloucester and Cheltenham tramway to gain access to Gloucester.

Construction started at Cheltenham, working northwards towards Birmingham, and the line was opened in stages, reaching Camp Hill (a temporary terminus in Birmingham) on 17 December 1840. The Railway was empowered by its Act to use the terminal station of the London and Birmingham Railway at Curzon Street (HEW 420, Chapter 6), to which it gained access via Gloucester Junction, running west over the L&B tracks to Curzon Street. The final link was opened on 17 August 1841, through trains running from Curzon Street to Gloucester from that date.

In 1854 the B&G trains were transferred to the newly built station at New Street. In 1864 an end-on junction was made between the B&G and the Birmingham and Derby Junction Railway, thus enabling through running between Derby and Gloucester.

The route of the B&G Railway follows the Severn valley northwards from Cheltenham for most of its length, but at Bromsgrove the railway is faced with a steep ascent up to the Birmingham area over the shoulder of the Lickey Hills. The great Lickey incline (HEW 865) was built two miles long at a gradient of 1 in 37.7 to lift the line over this obstacle. Opened in September 1840, the incline is straight with double tracks. It starts just north of Bromsgrove station (SO 969 693) and climbs to the summit near Blackwell (SO 991 718).

It was designed from the outset to be used by locomotive-hauled trains and in the early days of its operation special locomotives were imported from the United States of America, but these were replaced in 1845 by a more powerful locomotive, designed by James McConnell, and by others in succeeding years. As the weights of trains increased, the need for an even more powerful banking engine became more urgent and the Midland Railway (which had taken over the Birmingham and Gloucester Company in 1846) introduced a unique 0-10-0 locomotive designed by Sir Henry Fowler, which was in use from 1920 to 1956. Currently, modern diesel locomotives normally haul passenger trains up the incline unaided; banking assistance is usually provided for freight trains.

Nearby is the equally famous flight of 30 locks on the Worcester and Birmingham Canal at Tardebigge (HEW 768).

In the churchyard at Bromsgrove are the graves of two employees of the B&G Railway who lost their lives in an accident on the Lickey Incline in 1840. The tombstones depict locomotives of the period.

Today the Birmingham and Gloucester line is part of the North-East/South-West route in the British Rail network.

---

19.   TARDEBIGGE LOCKS   HEW 768
SO 963 680 to SO 995 693

Near the Lickey Incline (HEW 865), the Worcester and Birmingham Canal also faced a sharp climb from Bromsgrove towards Birmingham. Although gradients on the various roads in the vicinity are not abnormally steep, it is interesting that both canal and railway had to incorporate exceptional features which are widely known throughout the British Isles.

At Tardebigge between Redditch and Bromsgrove is the longest flight of locks in Britain, thirty in $2\frac{1}{2}$ miles giving a rise of 217 ft. These are numbered 29 to 58 and are closely preceded by another six at Stoke Prior, numbers 23 to 28, with a rise of 42 ft, and by yet another six at Astwood, numbers 17 to 22 – another 42 ft.

The top lock has an unusually large rise of 14 ft. It was originally the site of a vertical lift installed in 1808 and replaced by a conventional lock in 1815.

20. STOURPORT-ON-SEVERN **CANAL TERMINUS**	**HEW 241** SO 811 711

James Brindley sited the terminal basins of the Staffordshire and Worcestershire Canal (HEW 1083, Chapter 6) on a terrace of land about 30 ft above the Severn, well above any danger from floods. Originally the canal entered the Upper Basin (the more easterly of the high level basins) which communicated with the river through wide barge locks. This development was opened in 1770. A new basin to the west of the Upper Basin, and connected to the river by a set of narrow locks, opened in 1781.

Between these two basins is the Clock Warehouse, prominent among the original canal buildings, including the Tontine Inn, dating from the late 18th and early 19th centuries.

The Warehouse is a red brick two-storey building, 126 ft long and 25 ft wide with a slate roof. At the south-west corner a single storey lean-to has been built on, evidently a later addition as its roof cuts the first floor window openings. At the middle of the roof ridge a decorative timber cupola houses a four-faced clock.

The precise date of construction of the building is not known, but it is recorded that the clock was installed in 1812. Originally used as a store for grain and general goods, and later as part of a timber yard, it now houses the headquarters of the Stourport Yacht Club.

21. STOUR VALLEY HYDRAULIC ARCH	**HEW 1196** SO 83 79

The Stour Valley hydraulic arch forms part of a complex crossing of the River Stour and the Staffordshire and Worcestershire Canal by the Elan Valley water supply pipeline (HEW 1194, Chapter 7).

Three steel pipes are carried over the canal near the B4199 road about two miles north of Kidderminster, in the form of a three-centre segmental arch, completed in 1904, with a span of 106 ft 7 in. and a rise of 13 ft 8 in.

CHAPTER 6

West Midlands and South Staffordshire

Geographically, much of this region lies squarely on the watershed of England and thus there are few large natural rivers. In the north rises the high ground of Cannock Chase, while to the south the Birmingham plateau is bordered by the Clent and Lickey hills before descending into the valley of the Severn and its tributaries.

Economically, it is a region dominated by the intensively developed area between Birmingham and Wolverhampton, known colloquially as the 'Black Country'. Emerging rapidly in the late 18th and the 19th centuries, this region soon became a major centre of industry. Its rapid growth was mainly concerned with the metal trades: nail-making and chainmaking, locksmithing and gunsmithing, and the production of brassware, screws, nuts and bolts together with heavy engineering of high quality. Along with industry developed a demand for transport – always a necessary adjunct to the growth of factory-based manufacture. Partly owing to the lack of suitable natural waterways, canals were developed in the area to an extent not seen elsewhere and much of the network remains in use today. Birmingham was also served by two of the first major trunk railways, the London and Birmingham and the Grand Junction, and much early railway development took place here.

Although roads did not play a major part in the early development of the region, a notable exception is the Holyhead Road, improved under the guidance of Thomas Telford in the early 19th century, which passes through the area by way of Birmingham and Wolverhampton. More recently, the area has been the scene of much road building, being traversed by the M5 and M6 motorways with spectacular junctions at Gravelly Hill and Ray Hall.

Many of the early civil engineers were active in the provision of transport facilities in the region and some familiar names will be found in the following pages: Thomas Telford, James Brindley, Robert Stephenson and Joseph Locke, to name but a few. The well-known contractor, Thomas Brassey, started his railway

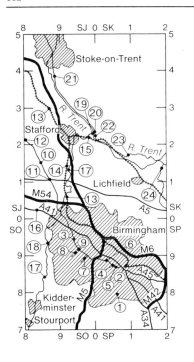

1. King's Norton Stop Lock
2. Curzon Street Railway Station, Birmingham
3. The Birmingham Canal Navigations
4. Engine Arm Aqueduct, Smethwick
5. Rabone Lane Canal Junction Bridges
6. Galton Bridge
7. Netherton Canal Tunnel
8. Dudley Canal Tunnel
9. Tipton Lift Bridge
10. Birmingham and Liverpool Junction Canal
11. Stretton Aqueduct
12. Shelmore Bank
13. Grand Junction Railway
14. Penkridge Viaduct
15. Shugborough Tunnel
16. The Holyhead Road
17. Staffordshire and Worcestershire Canal
18. Bratch Locks
19. Trent Aqueduct, Great Haywood
20. Great Haywood Canal Bridge
21. Trent and Mersey Canal
22. Essex Bridge, Great Haywood
23. Mavesyn Ridware Bridge
24. Chetwynd Bridge, Alrewas

building career here with the construction of Penkridge Viaduct on the Grand Junction Railway.

   To the north of the region a more rural landscape predominates in contrast with the 'dark satanic mills' of the Black Country. Even here the civil engineer has left his mark with the Trent & Mersey and Staffordshire & Worcestershire Canals, and the Grand Junction Railway and Trent Valley line which formed important links with the developing industrial areas of the north of England.

---

1.   KING'S NORTON STOP LOCK   HEW 975
                                                SP 056 759

The Stratford-upon-Avon Canal was authorized by an Act of Parliament of 1793 to run from a junction with the Worcester and Birmingham Canal at King's Norton to Stratford-upon-Avon. As the level of the water in the Stratford Canal was higher than the level in the Worcester and Birmingham Canal, a stop lock was built at King's Norton to prevent uncontrolled transfer of water from one to the other.

The lock is remarkable in having vertical 'guillotine' gates instead of the usual vertically-pivoted leaf lock gates. They are built in timber with a metal frame and are raised and lowered in a cast iron guide frame.

The movement of the gates is controlled by a winch sited on the north side of the lock at each gate. A lifting chain runs over a large pulley wheel mounted on the gate framework and round a single sheave pulley attached to the top of the gate. A counterbalance weight is suspended in a 'well' on the south side of the lock and is attached to the top of the gate by a chain which runs over two large pulley wheels. One of the advantages of the guillotine gate where the water level difference is small is that it removes the necessity for paddle gear.

The water level in the Worcester and Birmingham Canal was eventually raised to that of the Stratford canal, thus making the stop lock unnecessary, so the gates are permanently raised to permit uninterrupted passage of boats.

The Engineer to the Canal Company at the time was Josiah Clowes.

---

2. CURZON STREET RAILWAY STATION, BIRMINGHAM

HEW 420
SP 078 871

Illustrated on page 145

Robert Stephenson's standard gauge London and Birmingham Railway (HEW 1092) was, at 112 miles, the longest continuous railway so far built, and is generally accepted, together with the Stockton and Darlington (HEW 85)[1] and the Liverpool and Manchester (HEW 223)[2] railways, as being one of the three pioneering real public railways – all three the work of the Stephensons. The London and Birmingham reached Birmingham on 24 June 1838, the whole line from London having involved heavy engineering works through difficult country. Stephenson's assistant for the section from Rugby to Birmingham had been Thomas Gooch, brother of Brunel's Locomotive Superintendent on the Great Western Railway.

The terminus was not at the present New Street Station, but about half a mile to the north-east at the junction of Curzon Street and New Canal Street.

The station, opened in 1842, was designed by Philip Hardwick, architect of the original Euston Station, now vanished together with its great Doric arch, but some of whose features are reflected in its counterpart at the Birmingham end of the line. The squarish three-storey main station block which remains today is fronted by four giant Ionic columns and was originally flanked by archways leading into the station. The first floor windows are balustraded. The rear of the building has two further columns between outer square pilasters. The inside contains what was originally the booking hall with a steep iron balustraded stone stair, a refreshment room and offices. The station was built at a cost of £26 000.

The station was known as 'Birmingham' until November 1852 when the suffix 'Curzon Street' was adopted to distinguish it from 'New Street' and the newly opened Great Western Railway Station at Snow Hill. In July 1854 all regular trains were transferred to New Street which was reached by a short line from Curzon Street, also engineered by Robert Stephenson.

Thereafter, apart from holiday excursion trains which ran until 1893, Curzon Street handled goods traffic only. Today, although under threat of demolition, it still displays the label 'Goods Office' over the main doorway.

It was at the Queen's Hotel – not the present hotel of that name but its predecessor, which adjoined Curzon Street Station – that the Institution of Mechanical Engineers was founded on 27 January 1847 with George Stephenson as its first President.

3. THE BIRMINGHAM CANAL NAVIGATIONS HEW 1182

Illustrated on page 146

No review of the civil engineering heritage of Britain would be complete without a reference to the canal system which developed in Birmingham and the Black Country between 1768 and the 1830s. In this area the use of the canal for commercial transport was developed to a degree unsurpassed anywhere in the country.

The first canal in the area, built by the Birmingham Canal Company and opened in 1772, ran $22\frac{1}{2}$ miles from a junction with the Staffordshire and Worcestershire Canal (HEW 1083) at Aldersley west of Wolverhampton to wharves in what is now central Birmingham near Broad Street and Summer Row. It ran at three distinct 'levels': 453 ft from Birmingham to Smethwick, 491 ft over high ground at Smethwick, and 473 ft from Smethwick to Wolverhampton. The levels were connected by flights of locks and at Wolverhampton the canal descended to its junction with the Staffordshire and Worcestershire Canal by a flight of twenty locks

Curzon Street Railway Station: Ionic portico (Bourne)

(later increased to twenty-one). The Engineer for the project was James Brindley.

Later, major improvements were carried out along the line of the canal, notably the lowering of the summit level at Smethwick, first to 473 ft (1789) and subsequently by means of Telford's great cutting to 453 ft (1829); the Deepfields to Tipton cut-off through Coseley tunnel (1837); and the construction of a new line of canal between Smethwick and Tipton (1838).

The Birmingham Canal formed the focus of an intensive development of narrow canals promoted by the Birmingham Canal Company and other companies, which included the Dudley Canal, the Birmingham and Fazeley Canal, the Wyrley and Essington Canal, the Tame Valley Canal, the Rushall Canal and others. In addition, a large number of branch canals were built, serving the industry which grew up along their banks. Most of the canals served local needs but the Birmingham network was connected to the 'national' canal system by the Staffordshire and Worcestershire, the Warwick and Birmingham (later part of the Grand Union), the Stratford-upon-Avon, the Worcester and Birmingham and the Coventry Canals. In total, about 200 miles of canal were built in the area, more than in the city of Venice.

Water pumped from collieries was an important source of make-up water in parts of the network. Some of the old system has been abandoned and left to decay but much still survives and remains in use. Recently, some efforts have been made to improve the canal-

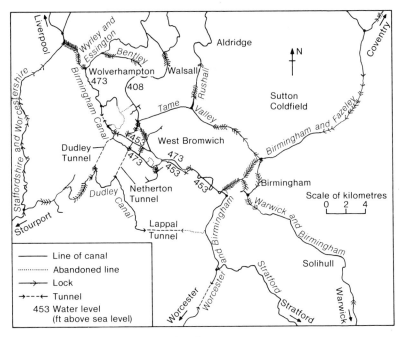

The Birmingham Canal Navigations

side environment in central Birmingham – a good example being the James Brindley Walk alongside the top of Farmers Bridge Locks on the site of the Newhall Branch canal, landscaping and other works having been carried out in 1968.

4.	ENGINE ARM AQUEDUCT,	HEW 492
	SMETHWICK	SP 024 888

This cast iron aqueduct carries the Rotton Park feeder at the 473 ft level of the Birmingham Canal over the later, lower line constructed in 1829. The difference between the water levels in the two canals is 20 ft.

Designed by Thomas Telford with the ironwork cast by the local Horseley Company the structure consists of a cast iron trough supported by a single arch of 52 ft span with five ribs, each comprising four sections with bolted joints. The spandrel bracings to the ribs are of the radial type intersected by a continuous arched member. The trough is supported on three of the ribs; the towpaths on either side are supported by cast iron arcades of unashamedly Gothic ecclesiastical-style arches and columns.

The waterway is 8 ft wide and each towpath is 4 ft 4 in. wide, that on the east side being paved in brickwork with raised strips to afford better grip for the horses' hooves.

| 5.   RABONE LANE CANAL JUNCTION | HEW 1197 |
| BRIDGES | SP 029 890 |

Illustrated on page 149

These two bridges are typical of cast iron towpath bridges found in many places on the canal network of Birmingham and the Black Country.

They were built in 1828 as part of Telford's improvements to the Birmingham Canal and connect various towpaths at the junction of the old and new lines.

Each bridge has a span of 52 ft 6 in. and a rise of 9 ft. Their semi-elliptical form has an advantage over the equivalent segmental arch in this situation as it gives greater headroom near the abutments where the bridge passes over a towpath.

The two ribs of each bridge were cast by the Horseley Ironworks and form the parapets, each in two sections. They have an X-lattice structure with a decorative quatrefoil pattern below the handrail.

The deck is of cast iron plates with raised ribs $2\frac{1}{2}$ in. high cast on their upper surface to help retain the earth filling which forms the footway.

The bridge abutments and wing walls are of brickwork with stone facing at the corners.

The ironwork of the bridges shows deep cuts caused by the abrasion of tow-ropes.

| 6.   GALTON BRIDGE | HEW 421 |
| Illustrated on page 150 | SP 015 893 |

When in 1829 the Birmingham Canal Company engaged Thomas Telford to improve the line of their canal, he avoided the six summit locks at Smethwick by excavating a great new cutting which took the canal at a lower level than the old line.

To carry the road from Smethwick to Sandwell across the cutting, Telford designed Galton Bridge, a single cast iron arch of

150 ft span, which springs from brick abutments set high up on the sides of the cutting. There are six arch ribs, each made up from seven segments with bolted joints. The space between the arch and the horizontal road deck is filled with diagonal intersecting ribs. The details resemble those of the Mythe (HEW 134, Chapter 5) and other Telford bridges.

Named after a member of the Birmingham Canal Committee, Galton Bridge is now relieved of its burden of traffic by the construction of a new road running parallel. It is interesting to note that the designers have avoided the necessity for a large bridge by putting the canal through a concrete 'tunnel' and constructing an embankment to carry the roadway.

---

7. NETHERTON CANAL TUNNEL   HEW 669
SO 954 883 to SO 967 908

The last major canal tunnel to be built in England, the Netherton Tunnel was opened in August 1858, construction having started at the end of 1855. It was built parallel to and about $1\frac{1}{2}$ miles to the east of Dudley Tunnel (HEW 670), in order to relieve traffic congestion on the Dudley Canal.

The tunnel is 1 mile 1267 yd long and is exceptionally large, being 27 ft wide, including two towpaths, and 16 ft high above the water level. When opened it was lit by gas, later converted to electric lighting.

The construction cost £302 000 compared with an estimate of £238 000, the difference being partly due to extra works necessitated by the condition of the ground through which the tunnel passes.

In 1983 the British Waterways Board replaced with concrete some 80 yd of brick invert which had heaved upward to such an extent as to impede traffic,[3] and restored to the tunnel the distinction of being the longest navigable on a canal route in Britain.

---

8. DUDLEY CANAL TUNNEL   HEW 670
SO 933 892 to SO 947 917

Early in the canal building age, Lord Dudley and Ward built a short canal, completed in 1778, to link his underground limestone quarries at Castle Hill, Dudley, to the Birmingham Canal. Later this canal was linked to the Dudley Canal (also promoted by Lord

Dudley) by means of a tunnel under the intervening ridge of high ground. After several setbacks the tunnel was finally opened in 1792.

At a length of 1 mile 1412 yd it was the fifth longest canal tunnel built in England. In so far as it also connected with underground mine workings, the tunnel has similarities with the Duke of Bridgewater's pioneering canal at Worsley (HEW 976, SD 749 005).[4]

Later tunnels were driven from the north end of the main tunnel to gain access to other limestone workings, in particular the Wrens Nest branch which was 1185 yd long and terminated in an underground basin.

The main tunnel is not continuous, for at its northern end it passes through two basins open to the air, the Shirts Mill and Castle Mill basins.

The size of the tunnel varies, reflecting the various stages of its construction and the effects of mining subsidence. A short section at the south end was rebuilt in 1884 to a larger size. At its deepest point the tunnel is about 200 ft below the surface. Most of the length is lined with brickwork but there are some unlined sections through harder rocks.

The tunnel was closed to traffic in 1962, but thanks to the efforts of a group of enthusiasts it has now been restored, and was re-opened in 1973.

Rabone Lane Canal Junction Bridges

9.   TIPTON LIFT BRIDGE   HEW 1225
SO 948 918

This vertical lift bridge, of unusual design, carried a road across an arm of Proof House Basin on the Birmingham Canal. It has a deck length of 27 ft 6 in. and a roadway width of 12 ft. The deck is lifted by four large chains, one being attached to each end of the two steel side plate girders. The chains pass over large wheels mounted at the top of vertical I-section steel columns and are connected to six-ton counterbalance weights at each end of the bridge.

The drive to the chain wheels is taken from a hand winch by a complex system of vertical and horizontal shafts connected by pairs of bevelled gears and a short length of chain drive. The vertical columns are cross-connected at the top by steel lattice girders.

The bridge was fabricated by Armstrong Whitworth & Co. in 1922 for the Great Western Railway. When the basin was closed in 1954, the bridge at that time was damaged and incapable of being lifted. After its removal from the basin it was eventually brought to the Black Country Museum at Dudley only a short distance away, where it has been repaired and re-erected across an arm of the Museum's canal system.

There is another lift bridge at Huddersfield (HEW 1307, SE 150 655), known as the Turn Bridge, as there was a swing bridge there until 1865.

Galton Bridge

10. BIRMINGHAM AND LIVERPOOL   HEW 1203
JUNCTION CANAL   SJ 902 020 to SJ 626 553

The Birmingham and Liverpool Junction Canal, opened in 1835 to connect the Birmingham Canal Navigations via the Staffordshire and Worcestershire Canal (HEW 1083) at Autherley to the Chester Canal north of Nantwich, a distance of 39½ miles, provided an alternative route to the Trent and Mersey Canal (HEW 1135) for trade between the Mersey and the West Midlands. There was also a 10¼ miles branch, with 23 locks, from Norbury Junction (SJ 793 227) to Wappenshall (SJ 663 147) on the Shrewsbury Canal, which linked East Shropshire with the Midlands. The main contractors were John Wilson and Sons.

The canal was constructed later than most. It was one of Thomas Telford's last works, and its style of construction suggests that its Engineer was endeavouring to compete with the up-and-coming railways.

The B&LJ Canal pursues a direct course across the landscape, rejecting the 'contouring' approach of earlier canal builders and making use of deep cuttings and high embankments to give long stretches free of locks. Where locks do occur, they are generally grouped together in flights, as at Audlem (15 locks). On the main line there are 28 locks and a stop-lock at Autherley, a short tunnel and a number of aqueducts.

A number of deviations from the originally planned line had to be made during construction because of objections by landowners. Some of these deviations necessitated extra works, most notably a bank and aqueduct at Nantwich with a revision to the original end-on junction with the Chester Canal and the embankment at Shelmore Wood (HEW 1224).

The canal was built in three sections and the first contract started in 1827. The canal was mainly completed by July 1833 but through traffic did not commence until 1835.

In 1842 the canal company introduced steam tugs to tow boats in trains, but they were considered to be uneconomic and horses were re-introduced.

Although Telford was the Engineer, William Cubitt deputized for him after February 1833 when Telford became ill. Telford died in September 1834, shortly before final completion. By amalgamation, the canal became part of the Shropshire Union in 1846. Commercial traffic ceased in the 1960s, but the canal is still busy with pleasure craft.

11. STRETTON AQUEDUCT   HEW 228
Illustrated on facing page   SJ 873 107

Constructed in 1832–33, Stretton Aqueduct carries the Birmingham and Liverpool Junction Canal in a cast iron trough over the A5 south-west of Penkridge in Staffordshire.

The trough is formed of five sections, each 6 ft 6 in. long, bolted together and supported by six cast iron arch ribs, each cast in two sections and joined at the centre of the arch. The clear span across the road is 30 ft, but the arch ribs are longer because the aqueduct crosses the road at a skew angle. The structure is 21 ft wide, the waterway narrowing to 11 ft across the aqueduct. Towpaths are provided on both sides, but only the one on the east side is currently in use.

On the east side is a cast iron parapet rail 3 ft 9 in. high, and each corner of the main arch is surmounted by an ornamental circular stone pillar. The abutments are in brickwork, curved in plan, and they include further stone columns matching those at the ends of the arch.

The designer, Telford, and date of the aqueduct are commemorated by an inscription on the centre panel of the trough. The builder was William Hazledine of Shrewsbury.

During 1961–62, the carriageway of the road under the aqueduct was lowered by about 4 ft to increase the headroom. Extra abutments were built for the aqueduct, with partial underpinning and support provided by an inverted portal-type structure beneath the new roadway.

12. SHELMORE BANK   HEW 1224
                                    SJ 805 215 to SJ 793 228

During the construction of the Birmingham and Liverpool Canal, it was necessary, because of opposition from Lord Anson, owner of Norbury Park, to alter the line of the canal so that instead of passing at ground level through Shelmore Wood it took a curved alignment across lower ground to the west. This deviation necessitated the construction of a massive embankment 1900 yd long and up to 60 ft high. Major problems were encountered during the forming of the embankment, with both settlement and slipping of the tipped soil. Eventually, after six years' work (1829–35), the bank was finished at the cost of delaying the complete opening of the canal by two years. Stop gates are provided at both ends of the bank.

Today thick vegetation covers the sides of the bank but the flat side slopes betray the difficulties encountered in stabilizing the fill. There are two road 'tunnels' under the bank with semi-circular masonry arches springing from vertical side walls. These may be compared with the tunnels at Beaminster (HEW 899, ST 468 032) or Charmouth (HEW 980, SY 349 948).

| 13. | **GRAND JUNCTION RAILWAY** | **HEW 1129** SP 078 871 to SJ 728 500 |

The earliest trunk railway route,[5] the Grand Junction Railway was opened on 4 July 1837, having been authorized by an Act of Parliament in 1833.[6] The railway ran for $82\frac{1}{2}$ miles from a junction with the newly-opened Liverpool and Manchester Railway (HEW 223)[7] at Newton-le-Willows near Warrington to its terminus in Birmingham at Curzon Street. It thus linked the expanding industrial areas of the Midlands to the industry and ports of the North-West.

The line was laid out by George Stephenson and John Rastrick, with Joseph Locke as assistant, but from the start of construction

Stretton Aqueduct

Locke was effectively in complete charge of the works although he was not formally appointed Chief Engineer until August 1835. It was on this line that the contractor Thomas Brassey began his long association with railway construction.

In the area covered by this chapter there are two notable viaducts, that at Aston and the long curved viaduct in the Rea Valley which led to the Birmingham terminus.

Shugborough Tunnel: west portal

The Grand Junction absorbed its progenitor, the Liverpool and Manchester, in 1845 and a year later merged with its contemporary, the London and Birmingham Railway, and with the Manchester and Birmingham (which ran only from Manchester to Crewe) to form the heart of the great London and North Western Railway.

In 1847 the Trent Valley line, engineered by Robert Stephenson, George Bidder and Thomas Gooch, with Thomas Brassey as Contractor, was opened to connect the Grand Junction Railway at Stafford with the London and Birmingham at Rugby, thus allowing traffic between London and the North-West to avoid Birmingham and with a nine-mile saving in distance. Although the line was completed by June 1847 full services did not begin until December because of the necessity of reviewing the cast iron structures on the line following the collapse of the Dee bridge on the Chester and Holyhead Railway (HEW 1094, Chapter 8). Like the Grand Junction Railway, the Trent Valley line became part of the London and North Western Railway, and it is still part of the British Rail West Coast Main Line.

---

14. PENKRIDGE VIADUCT
HEW 520
SJ 921 145

Carrying the Grand Junction Railway over the River Penk and a minor road, Penkridge Viaduct was the first major railway work to be constructed by the eminent railway contractor, Thomas Brassey.

There are seven arches of 30 ft span and the total length is about 240 ft. The piers and arches are constructed in brickwork with a stone cornice. There is a brick parapet wall. The piers are thickened as they near the ground, giving them a somewhat unusual curved profile.

---

15. SHUGBOROUGH TUNNEL
Illustrated on facing page
HEW 1126
SJ 981 216 to SJ 988 216

Shugborough Tunnel, the largest engineering work on the Trent Valley line, is 774 yd long and carries a double line of railway under the flank of the Satnall Hills, through the grounds of Shugborough Hall. The tunnel has a semicircular arch springing from vertical walls, is brick-lined and is built on a curve. The portals are built in ashlar masonry and have battlemented towers and, at the west portal, flanking walls leading to further ornate towers.

---

16.   THE HOLYHEAD ROAD   HEW 1213
SK 200 020 to SJ 800 110

Watling Street, the Roman Road from London to Shrewsbury (Uriconium) is followed closely by the present-day A5 trunk road. A military road, it was beautifully constructed on direct alignments but with no regard for social and commercial interests.

The route chosen by Telford for the Holyhead Road across the area of this chapter left Watling Street at Weedon Bec (SP 632 599) and followed what is now the A45 through Coventry to Birmingham; the A41 through West Bromwich, Bilston and Wolverhampton, then via the A464 through Shifnal to rejoin the line of Watling Street just west of Wellington. In 1919 the route was greatly improved for motor transport when the A5 became the Holyhead Road and avoided the Birmingham conurbation.

Telford's methods of road construction[8] may be contrasted with those of the Romans (HEW 630, Chapter 5) and of J. L. McAdam (HEW 1061, Chapter 4).

First, he levelled and drained the line of the road, employing a maximum gradient of 1 in 30 whenever possible and laying cross drains every 100 yd along the road, connected with ditches on each side. On the prepared foundation was laid a pavement, 7 in. deep in the centre and 5 in. deep at the edges, of large stones placed by hand as closely as possible, with their broad ends downwards: the top stones were all no wider than three inches. The interstices were hand-packed with smaller stones. Then followed a further 6 in. of stones weighing not more than six ounces and passing through a $2\frac{1}{2}$ in. ring. The final surfacing was a $1\frac{1}{2}$ in. thickness of gravel which soon, under the passage of wheels and horses' hooves, became a binding layer. A cross-camber of about four inches was provided.

For lightly trafficked roads, Telford used the solid base pavement as before, but the tops of protruding stones were broken off to take a layer of stones broken to walnut size, followed by the gravel surfacing.

Telford's methods were more expensive, but resulted in a more durable road than those of McAdam. The Roman technique of incorporating mortar to strengthen the construction was of course available to Telford, but would have added even more to the cost.

It must be remembered that the Roman roads were built by their troops or by conscript labour, whereas Telford had to employ paid workpeople.

17.   STAFFORDSHIRE AND   HEW 1083
WORCESTERSHIRE CANAL   SJ 995 230 to SO 813 708

This is one of the original canals built in the latter half of the 18th century. The fact that it is still open to traffic serves to demonstrate that the early canals were built to satisfy a perceived need for water transport, in contrast to the later speculative canals many of which proved commercially unsuccessful.

The Staffordshire and Worcestershire is a narrow canal, designed for boats 70 ft long and 7 ft beam. It runs for 46 miles from the Trent and Mersey Canal at Great Haywood to the Severn at Stourport, a town which owes its being to the canal. Begun in 1768 at its south end, with Brindley as Engineer, it was completed in May 1772.

From just east of Stafford, which was reached in 1816 by a now derelict branch, the canal climbs 100 ft by 12 locks to reach its summit level at Gailey Lock (SJ 920 106), where it passes under the A5 trunk road. Just north of the end of the $10\frac{1}{2}$ mile long summit level at Compton Lock (SJ 884 989) are the junctions with Telford's 1835 Birmingham and Liverpool Junction Canal (HEW 1203) at Autherley (SJ 902 020) and with the Birmingham Canal network at Aldersley (SJ 903 011). Until 1792 this afforded the only communication between Birmingham and the Severn. From the summit the canal begins a long descent by 29 locks to the valley of the River Stour, and follows the river closely to the Severn 265 ft below the summit.

18.   BRATCH LOCKS   HEW 982
SO 867 939

At Bratch, near Wombourne on the Staffordshire and Worcestershire Canal, is an interesting flight of three locks which between them lower the canal by 31 ft 2 in. in the course of its descent from the summit level into the valley of the River Stour.

The locks were designed by James Brindley and opened to traffic on 1 April 1771. Originally they were built as a staircase, single deep gates separating the locks. Later they were converted to three separate locks by the addition of top gates to the two lowest. The length of the intermediate 'pounds' between the locks was thus extremely short, about 15 ft. Such a short length of canal could not accommodate the water discharged from a lock without overflowing, so two long side ponds were constructed, connected to the canal by culverts through the canal walls. The side ponds are on the west side of the canal, the upper one extending for a considerable distance along the hillside.

The lock chambers, which are 70 ft long and 8 ft wide, reveal evidence of the alterations. The stonework of Bridge 48 was cut to allow the new top gate of the middle lock to open fully and the chamber walls of the middle and lower locks were recessed to receive the top gates when closed.

At the top of the locks is a neat octagonal toll house, strategically placed to observe traffic approaching from both directions.

Under the arch of Bratch Bridge across the lower lock is an unusually complex set of access stairways in brickwork.

---

19. **TRENT AQUEDUCT, GREAT HAYWOOD**	**HEW 885** SJ 994 229

Just south of Great Haywood Canal Bridge (HEW 676) this four-span stone aqueduct carries the Staffordshire and Worcestershire Canal over the River Trent. It was probably one of the later works to be built on the canal.

Brindley's technique in building works of this kind was to construct one or more arches on dry land, divert the flow of the river through them and then complete the aqueduct in the river bed.

Between the aqueduct and the canal bridge is a single arch over the site of the tailrace from the nearby mill.

There is another massive stone aqueduct (HEW 974, SJ 971 215) carrying the canal over the River Sow at Milford.

---

20. **GREAT HAYWOOD CANAL BRIDGE** Illustrated on facing page	**HEW 676** SJ 995 230

This elegant segmental arch, with a very flat span of 35 ft 4 in., carries the towpath of the Trent and Mersey Canal over the junction with the Staffordshire and Worcestershire Canal. The 20 in. thick arch ring is made up of two courses of brickwork with, over them, a 6 in. deep course of sandstone blocks.

Designed by James Brindley, the bridge was built about the year 1772.

21. TRENT AND MERSEY CANAL

HEW 1135
SK 200 180 to SJ 857 500

Among the earliest canals in Britain were those linking the four great rivers of central and southern England, the Mersey, Trent, Severn and Thames. One of the first was the 93 miles long Trent and Mersey Canal, also known as the Grand Trunk Canal and inspired by Joseph Wedgwood's concern for the transport difficulties experienced by himself and his fellow potters around Burslem and Stoke-on-Trent.

The canal, from the Trent at Wilden Ferry (SK 459 309) near the mouth of the Derbyshire River Derwent, runs through the area covered by this chapter via Stone and Stoke-on-Trent, rising 361 ft, with 40 locks and a short tunnel at Armitage (opened out in 1971), before passing under the high ground north-east of Stoke.

Work began in 1766 with James Brindley as Engineer, assisted by his clerk of works, Hugh Henshall, who took over on Brindley's untimely death in 1772. The canal was opened fully to traffic in May 1777 and continues largely unchanged to the present day,

Great Haywood Canal Bridge

although commercial traffic is now almost entirely replaced by leisure craft.

Apart from short lengths at each end the canal was designed for narrow boats, 70 ft long and with 6 ft 10 in. beam.

The canal connects with the national network at several places. At Great Haywood (SJ 995 230) it is joined by the Staffordshire and Worcestershire Canal (HEW 1083), and at Fradley (SK 140 140) near Lichfield a junction with the Coventry Canal (HEW 1090) affords access to the Thames via the Oxford Canal.

22. ESSEX BRIDGE, GREAT HAYWOOD   HEW 1132
                                        SJ 995 226

This ancient and historic sandstone masonry bridge, which spans the River Trent between Great Haywood and Shugborough Park, Staffordshire, has 14 arches of segmental form with voussoirs 14 in. deep. The total length is about 310 ft, the spans of the arches varying between 14 and 15 ft. The line of the bridge is generally straight but at its south end there is a marked curve to the east, one of the arches being curved in plan to fit. The bridge is only 5 ft wide with stone parapet walls, giving a footway width of about 4 ft 3 in.

At each pier there are triangular cut-waters on both sides of the bridge, and these are continued up to footway level to form pedestrian refuges 2 ft 3 in. deep and 4 ft wide.

The exact date of construction is unknown but the bridge is reputed to have been built at the end of the 16th century by the first Earl of Essex to carry his horses and hounds over the river on their way to Cannock Chase. It seems originally to have been much longer since in the late 17th century there was a reference to a bridge having 43 arches compared with the present total of only fourteen. It may be presumed that the bridge continued as a causeway across the low-lying land bordering the river.

23. MAVESYN RIDWARE BRIDGE   HEW 726
                                        SK 092 168

The 'High Bridge' carrying the B5014 over the River Trent between Mavesyn Ridware and Armitage is one of two substantial cast iron bridges built in east Staffordshire in the early 19th century under the direction of Joseph Potter, the County Surveyor.

The cast iron arch of 140 ft span was cast by the Coalbrookdale Company in 1830.

Each of the five ribs, 36 in. deep and 2 in. thick, has seven bolted segments. The arch is given lateral stability by diagonal bracing running horizontally between the ribs and by transverse connections between the ribs at each segment joint. The spandrel is of the X pattern made up from vertical and inclined members giving a 'latticed' appearance, enhanced by the arched member joining their intersections.

The roadway is 17 ft 9 in. wide and the total width of the bridge, including two narrow footpaths and the parapets, is 25 ft 8 in. There is a cast iron parapet fence 3 ft high. The stone abutments have wings which are curved in plan and incorporate massive stone pilasters.

In 1982 a major operation was undertaken at the expense of the National Coal Board to safeguard the bridge during an expected period of mining subsidence. The road was diverted on to a Bailey Bridge alongside. A steel arch was built on concrete piers in the river to take the weight of the old arch, which was lightened by removing the handrailing and the road pavement, while in each rib a joint near the crown was unbolted so that relative movement of the original bridge abutments could take place without fracturing the ribs.

24.   CHETWYND BRIDGE, ALREWAS   HEW 168
SK 187 139

The other cast iron bridge built under the direction of Joseph Potter carries the A513 road over the River Tame. It was built in 1824 and probably cast at Coalbrookdale. The three arched spans are 71 ft 6 in., 81 ft 3 in. and 72 ft 5 in. Each arch has five ribs each of five segments with X-type spandrel bracing having a continuous arch-like member joining the intersections. The roadway is 18 ft wide with a narrow footpath.

CHAPTER 7

Shropshire

Although now generally regarded as being of a rural character, Shropshire, for a relatively short time, was one of the cradles of the industrial revolution. Along the gorge of the upper Severn was found the necessary combination of raw materials (coal and iron ore), transport (the river), and inventiveness (Abraham Darby and his successors), which led to the development of the historic ironfounding industry of the area.

A glance at the map will show from the distribution of the civil engineering works which are described in the following pages the major influence of the Severn Valley on industrial development. The river has acted as a major trade artery, attracting to its banks

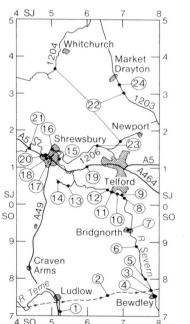

1. Ashford Carbonel Bridge
2. Elan Valley Aqueduct
3. Elan Valley Pipeline Bridge, Bewdley
4. Bewdley Bridge
5. Victoria Bridge, Arley
6. Mor Brook Bridge
7. St Mary Magdalene Church, Bridgnorth
8. Coalport Bridge
9. Hay (Coalport) Inclined Plane
10. The Free Bridge, Jackfield
11. The Iron Bridge, Coalbrookdale
12. Albert Edward Bridge, near Buildwas
13. Cound Arbour Bridge
14. Cantlop Bridge
15. Belvidere Bridge, Shrewsbury
16. Bage's Mill, Shrewsbury
17. Castle Walk Footbridge, Shrewsbury
18. Coleham Pumping Station, Shrewsbury
19. The Holyhead Road
20. English and Welsh Bridges, Shrewsbury
21. Montford Bridge
22. Shropshire Union Canal
23. Longdon-upon-Tern Aqueduct
24. Tyrley Cutting

many industrial developments, but it has also formed a major barrier to communication from east to west, so that many of the works are concerned with crossings of the river by both road and rail.

Although many of the early civil engineers worked in the area, the name of Thomas Telford deserves particular mention. Early in his career, first as an architect and then as a major figure in civil engineering, he worked extensively in Shropshire and held several appointments including that of Surveyor of Public Works for the County from 1787.

The more recent history of the region shows a marked reduction in its industrial importance. The coal and iron industry declined in the 19th century, whereas the neighbouring Black Country between Birmingham and Wolverhampton continued to prosper. The establishment of the New Town, appropriately named after Telford, in the 1960s has helped not only to revitalize this area but also to encourage the preservation of many of the historic works described in this book.

1. ASHFORD CARBONEL BRIDGE   HEW 1101
SO 520 711

A fine single-span, segmental masonry arch bridge carries the minor road from Ashford Carbonel to Ashford Bowdler over the River Teme south of Ludlow. Begun in 1795 and opened on 25 November 1797, the arch has a span of 81 ft and a rise of 24 ft 6 in. The voussoirs are mostly of gritstone with some limestone blocks. The spandrel walls are built in local red sandstone. The carriageway is 16 ft wide and there are no footpaths. A particular feature of the bridge is the paved invert between the abutments below river level.

The bridge was designed by Thomas Stainston for Thomas Telford when the latter was County Surveyor of Shropshire. A contract for its construction was drawn up between the Clerk of the Peace for the County and William Atkins, stonemason, and Thomas Smith, carpenter.

2. ELAN VALLEY AQUEDUCT   HEW 1194
SN 93 65 to SP 00 80

The aqueduct carries water from a draw-off at Caban Coch reservoir (HEW 550, Chapter 2) a total distance of 73 miles to a storage reservoir at Frankley to the south-west of the City of Birmingham.[1] Since this reservoir is about 170 ft lower than the inlet to the aqueduct at Caban Coch, the water is able to flow under gravity for the whole distance.

The general route is via Rhayader, Knighton, Ludlow, Cleobury Mortimer, Bewdley and Hagley to Frankley. For half its length the aqueduct acts as a conduit (that is, the water flows at atmospheric pressure) and for the remainder the water flows in cast iron or steel pipes under pressure.

In the initial scheme, first used in 1904 and officially opened in 1906, two pipes were installed, and two more followed between 1919 and 1961.

In the construction of the aqueduct, fifteen tunnels were driven with a total length of 12.9 miles. The longest (Dolau Tunnel) is $4\frac{1}{4}$ miles long.[2] There are also several major structures which carry the aqueduct across roads, rivers, railways and other obstacles. Among the more notable are: crossings of the River Teme at Leintwardine, Downton-on-the-Rock and Ludlow; the River Severn crossing at Bewdley (HEW 1195); the Stour Valley crossing of the River Stour and the Staffordshire and Worcestershire Canal (HEW 1196, Chapter 5); and Hagley railway crossing.

Victoria Bridge, Arley (Severn Valley Railway)

3. ELAN VALLEY PIPELINE   HEW 1195
BRIDGE, BEWDLEY   SO 77 78

The Elan Valley Aqueduct crosses the River Severn, two miles north of Bewdley, on an impressive segmental steel arch of 150 ft span and 15 ft rise.[3] There are four arch ribs set at 12 ft 6 in. centres, each made up of rivetted steelwork with top and bottom flanges separated by an 'X' lattice. At the abutments each arch rib rests on an 8 in. dia. steel pin with a cast iron shoe and heel plate bolted to a granite block.

There are three pipe galleries originally designed to take two 42 in. dia. pipes in each. This is the lowest point on the aqueduct and the water pressure is 250 lb/sq.in.

On the east bank of the river there is an approach viaduct over the flood plain with five segmental arches of spans varying from 34 ft to 62 ft built in brickwork with stone facings. The total length of the viaduct and bridge is 624 ft.

4. BEWDLEY BRIDGE   HEW 461
                                      SO 788 755

The present masonry bridge at Bewdley is one of several which have crossed the Severn here since 1447. The first was destroyed in 1459 by the Lancastrians and was replaced in 1460 by a timber bridge which was rebuilt in 1483. This in turn was damaged in the Civil War but repaired in 1644, and remained until it was swept away by the Severn flood of 1795.

The design for a new bridge was prepared by Thomas Telford. The summer and autumn of 1798 were very dry and the bridge was, astonishingly, built in that one period, the contractor being John Simpson of Shrewsbury. The three main arches carry the A456 Kidderminster to Ludlow road over the river. The voussoirs have deep V joints. The central arch has a span of 60 ft and a rise of 18 ft and the two side arches have spans of 52 ft with a rise of 16 ft 9 in. There are two smaller arches on each bank, which carry the road over the river towpath and allow an easier passage for floodwater. The overall width of the bridge is 27 ft. Its balustraded stone parapet is 3 ft 10 in. high.

The road to the east of the bridge runs parallel to the river bank on a low viaduct of twelve semi-circular arches surmounted by a neat cast iron balustrade.

5.   VICTORIA BRIDGE, ARLEY   HEW 464
Illustrated on page 164   SO 767 792

The Severn Valley Railway, as its name suggests, follows the narrow valley of the River Severn and crosses it just south of the village of Arley. At this point was built the Victoria Bridge, which at the time of its construction had the longest span, 200 ft, of any cast iron bridge in Britain. It was opened for traffic on 31 January 1861, and carries a single line of railway.

The arch is made up of four ribs, each consisting of nine bolted segments. An inscription cast into the centre of the arch records that it was designed by Sir John Fowler and cast by the Coalbrookdale Company. The contractors were Brassey, Peto and Betts of London.

The stone abutments of the bridge are pierced by brick arches allowing access along the banks of the river.

The line was closed to traffic in 1963 but the section from Bridgnorth to Bewdley including the Victoria Bridge has been re-opened by a private company. Passengers may still cross the bridge in steam-hauled trains, a fact which has not been overlooked by television and film producers.

6.   MOR BROOK BRIDGE   HEW 972
                                                     SO 733 885

From ancient times the Severn has been a trade route. In 1772 the towpath of the river from Bewdley to Coalbrookdale was, unusually, the subject of the creation of a Turnpike Trust. It was improved in the early 19th century to allow horses instead of manpower to be used for hauling boats.

Wherever a side stream entered the river, a bridge was necessary to carry the towpath over the obstruction. Mor Brook Bridge, about four miles south of Bridgnorth, on the west bank, was built in 1824. It has a single cast iron span with three ribs, each cast in two sections with a joint at the centre. The arch spans 30 ft between vertical brick abutments and the bridge is 5 ft 9 in. wide. Four-inch dia. tie rods are used to connect the arch ribs laterally. There is an attractive parapet of cast iron X-lattice form with a decorative handrail.

The ironwork for the bridge was cast by John Onions of Broseley.

7. ST MARY MAGDALENE CHURCH, BRIDGNORTH

HEW 1134
SO 717 928

Illustrated below

Thomas Telford, although best remembered as a civil engineer, had, in his earlier years, unrivalled experience as a master mason engaged on some of the finest monumental architecture of the time, such as Somerset House in London and Portsmouth Dockyard. Amongst several churches which he designed in the last years of the 18th century, one of the best examples is St Mary Magdalene, the parish church of Bridgnorth.

Prominently situated, the church has an unusual orientation, the nave lying north-south (rather than the traditional east-west). The tower is 120 ft high with a clock and eight bells and a copper covered roof. The interior of the church is simple in design, the almost square nave being divided by two rows of Ionic columns

St Mary Magdalene Church, Bridgnorth (Telford: *Life and atlas*)

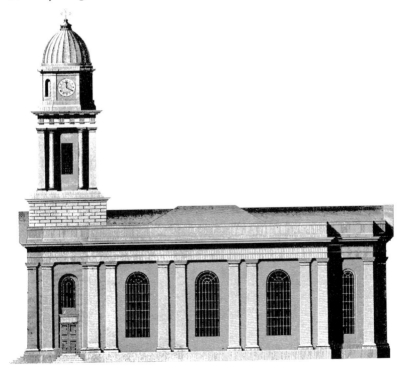

with a timber gallery at the north end. The large arched windows in plain glass, extending the full height of the side walls, were intended to create (in Telford's own words) the appearance of 'one great and undivided apartment'.[4]

The church was built between 1792 and 1795 by John Rhodes and Michael Head.

---

8. COALPORT BRIDGE

HEW 422
SJ 702 021

The 103 ft span cast iron bridge across the Severn, just below Coalport, was built by John Onions of Broseley in 1818 and is the third bridge on the site.[5] The first, built in 1780, had two timber arches with masonry abutments and central pier. In 1799 the pier was removed and the timber arches were replaced by a single span of three cast iron ribs to support the timber deck and parapets. The centre rib fractured in 1817. The present bridge includes some of the arch ribs from its predecessor, but with the number of its ribs increased to five. It is one of the oldest cast iron bridges still carrying traffic, and is a scheduled Ancient Monument.

---

9. HAY (COALPORT) INCLINED PLANE

HEW 639
SJ 694 026

The Hay Inclined Plane at Coalport, designed by William Reynolds, was built in 1792–93 to transport boats between the Shropshire Canal and the River Severn. The plane raised tub boats in the dry on trolleys running on two parallel standard gauge rail tracks, two boats being carried simultaneously, one on each track. As the main traffic of the Canal was down to the Severn, the heavier boat descending the plane drew up the lighter ascending boat, the two trolleys being linked by ropes to a common winding drum on which was a brake wheel. A steam beam engine at the upper level was used solely to lift the loaded trolley over the sill of the upper basin after which it descended by gravity.

The plane lifted the boats through a vertical height of 207 ft in a length of approximately 300 yd, a gradient of about 1 in 4. It could pass a pair of five-ton tub boats in $3\frac{1}{2}$ minutes instead of the $3\frac{1}{2}$ hours which would have been spent in passing locks, possibly as many as 28. The incline finally ceased to operate in 1907 and the rails were removed in 1910.

Shortly after the Ironbridge Gorge Museum Trust was formed in 1968, the upper basin and the incline were cleared. Restoration of the rail tracks and the lower basin followed in 1975–76.

10.   THE FREE BRIDGE, JACKFIELD   HEW 658
SJ 681 033

This bridge, in the relatively new material of reinforced concrete, was built in 1909 across the Severn, about half a mile below the Iron Bridge, to meet the demand for a toll-free crossing in the area.

The central span of 80 ft and the two side spans of 56 ft each have two segmental arched ribs joined by transverse horizontal beams. The deck is supported by vertical columns rising from the ribs.

The bridge was designed by L. G. Mouchel and Partners and built by the Liverpool Hennebique Company.

11.   THE IRON BRIDGE, COALBROOKDALE   HEW 136
Illustrated on page 171   SJ 672 034

'One of the boldest attempts with a new material was the application of cast iron to bridges' – Thomas Tredgold, 1842[6]

The upper Severn gorge became one of the birthplaces of modern industry when in 1709 Abraham Darby, originally a Bristol brass founder, began to smelt local iron ore with coke made from local coal at Coalbrookdale. To enhance the communications of the area it was proposed to construct a bridge over the River Severn between Broseley and Madeley Wood. Severe floods in earlier years suggested a single span to avoid piers in the river. In February 1776, a Bill was laid before Parliament for the construction of a bridge in cast iron. Thus was born the Iron Bridge, the first major bridge in the world to be constructed wholly of cast iron, and which gave its name, Ironbridge, to the settlement which sprang up around it.[7]

The bridge clears the river in a single arch of 100 ft span and is made from ten half-ribs, each cast in one piece by Abraham Darby III in his Coalbrookdale furnace. The project in essence is credited to the architect Thomas F. Pritchard. The detailed designs for the ribs and members of the bridge were prepared under the direction of Abraham Darby, grandson of the first owner of the foundry, by Thomas Gregory, foreman pattern maker at Coalbrookdale and therefore a worker in wood, and this probably explains the construction being in the timber style with mortise and tenon and dovetail joints.

Construction of the bridge took one and half years and it was opened on New Year's Day 1781. It contains just over 378 tons of ironwork, probably equivalent to three or four months output from a contemporary furnace. It survived flood and tempest to carry vehicular traffic until 1931 when it was closed to all but pedestrians. Major repairs were necessary at intervals during the life of the bridge, being mainly occasioned by the tendency of the sides of the gorge to move towards the river.

When, in 1796, Telford undertook a cast iron bridge at Buildwas, two miles upstream (HEW 648, SJ 645 045) he was aware of this tendency since he states 'I made the arch 130 ft span. The road rested on a very flat arch calculated to resist the abutments if disposed to slide inwards as at Coalbrookdale'. Alas, he was no more successful and the main arch broke in 1889 to be replaced by the present bridge in 1905.

In 1973, under the initiative of the Ironbridge Gorge Museum Trust, a reinforced concrete invert was constructed in the bed of the river between the two abutments of the Iron Bridge.[8]

The bridge and its adjacent tollhouse on the south side are now preserved as a monument to the pioneering spirit of British engineers and craftsmen.

| 12. | **ALBERT EDWARD BRIDGE,** | **HEW 350** |
| | **NEAR BUILDWAS** | SJ 659 038 |

Opened on 1 November 1864, the Albert Edward Bridge carries the double line of the Wenlock Railway, now part of the London Midland Region network, over the River Severn. It is similar to the Victoria Bridge near Arley (HEW 464) and like it was designed by Sir John Fowler and cast by the Coalbrookdale Company.

The four cast iron ribs are in nine sections bolted together and spring from brick abutments. The decking is supported from the arch by cast iron verticals heavily cross-braced. The original wrought iron and timber decking was replaced in 1933 by steel beams and plates, supporting ballasted track.

This is thought to be one of the last, if not the last, major cast iron railway bridges to have been built and is still in use today, carrying the daily coal supply to the Ironbridge Electricity Generating Station nearby.

13.   COUND ARBOUR BRIDGE   HEW 423
SJ 555 053

The oldest cast iron bridge in the county of Shropshire in use by modern traffic, Cound Arbour Bridge was built in 1797. It carries an unclassified road over the Cound Brook five and a half miles south-east of Shrewsbury.

The bridge spans 36 ft and comprises three cast iron ribs manufactured at Coalbrookdale. The outer two are decorated by circles filling the space between the upper and lower edges of the ribs. Cast iron plates carried directly on the ribs span the full 14 ft width. The space between the plates and the road surface was originally filled with loose material, retained by cast iron plates.

In 1920, this filling was replaced by mass concrete and in 1931 new mass concrete abutments were provided. Since then the concrete, by acting as an arch, has helped the original ribs to carry the load of modern traffic.

The Iron Bridge, Coalbrookdale (Ironbridge Gorge Museum Trust)

14.   CANTLOP BRIDGE

HEW 330
SJ 517 063

Cantlop Bridge, also across the Cound Brook, was built in the year 1812 with a single span of 32 ft. It is the only survivor of four similar cast iron bridges.

It is not known for certain who designed it. Telford, in his capacity as County Surveyor of Shropshire, stated in a report to the Sessions of July 1812 that 'the bridge at Cantlop Ford although not enlarged as first suggested by him (Telford) will according to the present plan be placed in such a footing as to be safely accepted as a County Bridge'. So well built was it that only small repairs and painting needed to be done during the next 163 years.

The bridge is scheduled as an Ancient Monument and in 1975 was by-passed to ensure its preservation. Ribs similar to those of Cantlop Bridge from bridges at Cound and Stokesay may be seen in Ironbridge and Coalbrookdale Museums.

15.   BELVIDERE BRIDGE, SHREWSBURY

HEW 903
Illustrated on facing page
SJ 520 125

In 1849 this bridge was built to carry the Shrewsbury and Birmingham Railway across the Severn just east of Shrewsbury.

Each of its two skew spans of 101 ft 6 in. (89 ft on the square) is a cast iron arch with six sectional ribs and a cast iron parapet. The rise of each arch is 10 ft 8 in. The abutments and 13 ft thick central pier are of masonry.

Apart from some renewal of deckplates, scour protection and grouting of the pier the bridge required little maintenance from its construction until 1984, when a reinforced concrete deck replaced the original cast iron deckplates. The bridge continues to carry passenger and freight trains with axle loads as high as 25 tons.

The arch sections were cast at Coalbrookdale, the Engineer was William Baker and the contractors were Hammond and Murray.

16.   BAGE'S MILL, SHREWSBURY

HEW 425
Illustrated on page 175
SJ 500 140

In 1796, work began on a flax mill in Castle Foregate on the northern outskirts of Shrewsbury. Powered by a 20 hp Boulton and

Watt engine, the mill was completed in August 1797 to the design of Charles Bage, a Shrewsbury wine merchant, amateur engineer and friend of Telford, whose letters to William Strutt of Derby contain the earliest analysis of the strength of iron beams and columns. He had been brought into partnership, for his technical knowledge, by the Shrewsbury flax merchants Thomas and Benjamin Benyon, who were already in partnership with John Marshall of Leeds.

The mill is 174 ft long and 36 ft wide internally, and is five storeys high. The floors are carried on brick arches springing from cast iron beams supported by the external brick walls and three rows of cruciform section tapered cast iron columns. Hazledine's Shrewsbury foundry supplied the ironwork.

Bage's mill is the world's first multi-storey building with an interior iron frame.[9] Its structure contained no combustible materials – an enormous advantage over the earlier type of textile mills with wooden floors. It was quickly followed by a sucession of iron-framed mills at Salford, Belper, Leeds and elsewhere with improved designs of the beams attributable largely to Bage himself. About the year 1897, the Shrewsbury mill was converted to a malthouse, but it still retains its original structure in excellent condition.

Belvidere Bridge

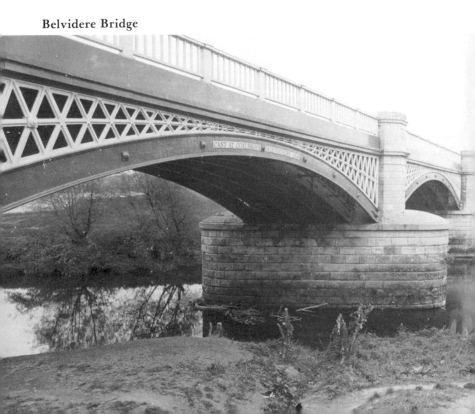

17.   CASTLE WALK FOOTBRIDGE,   HEW 1105
         SHREWSBURY   SJ 499 130

Castle Walk Footbridge, one of numerous river crossings in
Shrewsbury, is for pedestrians only and when opened in
November 1951 was the first bridge of its type to be built in Shropshire. It replaced a steel wire rope suspension bridge built in 1910
which had become unsafe owing to corrosion.

The bridge is formed from two balanced cantilevers, the ends of
which support a central span, and is built of post-tensioned
prestressed concrete. The central span is 150 ft and the two side
spans are each 33 ft. The footpath is 10 ft wide between the
parapets.

The bridge was designed by architects T. P. Bennet & Company
and consulting engineers L. G. Mouchel and Partners in association with the Prestressed Concrete Company. The builders were
Taylor Woodrow Construction.

18.   COLEHAM PUMPING STATION,   HEW 424
         SHREWSBURY   SJ 496 121
Illustrated on page 176

Shrewsbury allowed all its drainage to run direct into the Severn
until the 1890s when a series of main collecting sewers were constructed parallel to the river banks to intercept and pick up the foul
sewage. At Coleham, to the south-east of the City centre, the
sewers on the north bank were connected to a cast iron main sewer
running underneath the river to a new pumping station on the
south bank.

The brick-built station was equipped with twin beam pumping
engines, each engine being a two-cylinder steam compound built
by Messrs Renshaw. The two cylinders are vertical and are connected to one end of the overhead beam with a crankshaft and a
16 ft dia. flywheel at the other. Steam at a pressure of 90 lb/sq.in.
was supplied from two boilers. The speed of the engines was
15 rev./min., 114 gallons being pumped with each stroke.

The sewage was pumped through cast iron pipes to a point in
Whitehall Street from which it ran by gravity to the Sewage Treatment Works at Monkmoor. The whole system was opened on
1 January 1901.

The pumping engines continued in service until 1970, when
electric pumps were installed, but the engines have been preserved.

Bage's Mill (Ironbridge Gorge Museum Trust)

---

19.   THE HOLYHEAD ROAD   HEW 1212, 1213
SJ 800 032 to SJ 400 176

Westward from Oakengates, now part of the New Town of
Telford, Telford's Holyhead Road is followed by the present A5
trunk road except where improvements to the latter, such as the
by-pass to the south of Shrewsbury, have departed from the
original route.

The Holyhead Road crossed the Severn in Shrewsbury on the
English and Welsh Bridges (HEW 1026, 1025) and turned north-
west to follow the Dee valley through Chirk and Llangollen and so
on across Wales (Chapter 1).

The Holyhead Road Commissioners had agreed with the seventeen Turnpike Trusts between Shrewsbury and London that each
should retain its own surveyor who would be responsible to Telford
as General Surveyor. By 1829, he had supervised the rebuilding of
this 267 miles long trunk road over which coaches could now
travel at an average of 12 miles/h – only a few years before the
railways came to render such a speed a thing of the past.

Coleham Pumping Station (Ironbridge Gorge Museum Trust)

20. ENGLISH AND WELSH   HEW 1026, SJ 496 124
BRIDGES, SHREWSBURY   HEW 1025, SJ 489 127

Illustrated on page 179

The historic town of Shrewsbury is built on a hill in the middle of a large loop of the River Severn, which from earliest times has necessitated river crossings to east and west.

The modern A5 trunk road, in by-passing to the south, has departed from the original route of the Holyhead Road, which now is the A458 through the town.

The English Bridge is the eastern crossing. The first crossing here, in 1550, was a five-arch bridge, with a causeway of twelve culverts, which was replaced in 1774 by a new bridge designed by John Gwynn of Shrewsbury. This bridge, of Grinshill sandstone, was 400 ft long with seven semi-circular arches, the central arch having a span of 55 ft. The overall width was 23 ft 6 in. with a roadway only 15 ft wide.

In 1925 the complete rebuilding of Gwynn's bridge was begun. The old structure was demolished and the stones were re-dressed and re-used. The reconstructed bridge was given the same general configuration as its predecessor, with a centre span of 54 ft 6 in. flanked on each side by spans of 47 ft 3 in., 41 ft 9 in. and 36 ft 6 in. in succession. The height at the crown of the centre arch was reduced by 5 ft and the overall width was more than doubled to 50 ft between parapets, with a reinforced concrete core and deck. The reconstructed bridge was formally opened by HM Queen Mary on 26 October 1927.

The Welsh Bridge, on the west of the town, is the second on the site. The first was built about 1262 and was dismantled when the present bridge was opened, some 40 yd upstream, in 1794. It was designed and built by John Carline and John Tilley. Carline's father, also John, was foreman mason on the English Bridge. The overall length is 266 ft, with five sandstone arches. The span of the centre arch is 46 ft 2 in. and that of the others is 43 ft 4 in. The width between parapets is 30 ft.

At mid-span there is a fitting for a pulley wheel used for hauling boats over the shallows downstream of the bridge.

---

21. MONTFORD BRIDGE   HEW 333
                                                      SJ 432 153

The first bridge to be designed by Thomas Telford after he was appointed County Surveyor of Shropshire, Montford Bridge carries the A5 Holyhead Road over the River Severn four miles north-west of Shrewsbury. The bridge was built by John Carline (Junior) and John Tilley between 1790 and 1792, of local red sandstone quarried at Nesscliffe.

The bridge has three spans, the central span being 58 ft and the two outer spans 50 ft. The centre arch is 24 ft above mean water level. The original width of the roadway was 20 ft, but in 1963 the carriageway was widened to 22 ft and two footpaths each 4 ft 6 in. wide were provided. This was achieved by constructing a reinforced concrete slab, cantilevered 5 ft 6 in. beyond the existing width of the bridge on both sides.

To superintend the construction of Montford Bridge, Telford sent for his old friend and workmate Mathew Davidson of Langholm; he also employed a mason from Shrewsbury named John Simpson. This was the start of a lifelong association between the three men.

22.   SHROPSHIRE UNION CANAL   HEW 1202 – 1206

The Shropshire Union Canal system resulted from the merger, circa 1846, of several canals and canal-owning railways.

It may be divided into five major groups each with its various branches

- (a) The Ellesmere Canal, with the Chester Canal (HEW 1202, Chapter 8)
- (b) The Birmingham and Liverpool Junction Canal (HEW 1203, Chapter 6)
- (c) The Llangollen Branch of the Ellesmere Canal (HEW 1204, Chapter 1)
- (d) The Montgomeryshire Canal (HEW 1205)
- (e) The Shrewsbury Canal (HEW 1206).

In 1847 the London and North Western Railway Company leased the whole of the system. Since early in the 20th century there have been progressive closures of branches and sections of main line but although much has disappeared, much remains, even though it may be only used for pleasure craft, and restoration is actively proceeding in various places.

The main line of the Ellesmere Canal from Chester to Frankton was opened in 1805, as was the Llangollen branch. The Montgomeryshire Canal opened in 1797 from Carreghofa in Shropshire (SJ 255 203) on the Llangollen branch as far as Garthmyle (SO 194 994) south of Welshpool, with members of the Dadford family as Engineers, but it was not until 1819, under Josias Jessop, that it reached Newtown on the Severn (SO 113 916). There is now little left of this canal, but a pair of cast iron gates from Welshpool can be seen at the Waterways Museum, Stoke Bruerne, Northamptonshire.

The Shrewsbury Canal opened in 1796 over the 17 miles from Shrewsbury to Wellington – Oakengates (SJ 672 126). Josiah Clowes was succeeded as Engineer by Thomas Telford in 1795. The more important works included Longdon-upon-Tern Aqueduct (HEW 280), an inclined plane with a 75 ft lift at Trench and the 970 yd long Berwick Tunnel, the first tunnel to have a towpath, which took the shape of a 3 ft wide timber gangway on bearers set in the wall.

At Wellington, the Shrewsbury Canal joined the 1788 Shropshire Canal, which, with three inclined planes, led to the Severn at Coalport, some 22 miles below Shrewsbury.

English Bridge, Shrewsbury (Shropshire County Library)

23. LONGDON-UPON-TERN AQUEDUCT   HEW 280
Illustrated on page 180   SJ 617 156

The Longdon-upon-Tern aqueduct carries the Shrewsbury Canal
over the River Tern, a small tributary of the Severn, close to the
Wellington to Shawbury road, the B5063, about five miles north-
west of Wellington.

Built in 1796, it was the first canal aqueduct to be designed in
cast iron, and was completed about one month after Outram's
aqueduct (demolished in 1971) on the Derby Canal. It has a trough
9 ft wide and 3 ft deep, and is 186 ft long with four spans of 47 ft
8 in. The slender lines of the trough and its three-legged vertical
and inclined cruciform cast iron supports are in strange contrast
with the heavy masonry abutments (each with two flood arches).
The reason for this may be found in the Shrewsbury Canal Company's minute book which records that Mr Clowes, the Engineer,
had died in 1795 and that Thomas Telford had been appointed to
succeed him. It would seem that Clowes had already completed a
masonry aqueduct which was either unsatisfactory to Telford or
had been damaged by floods, and this new form of iron structure
replaced it. It is interesting to note that Outram was Jessop's part-

Longdon-upon-Tern Aqueduct

ner in the Butterley Ironworks and that Telford had worked under Jessop on the Ellesmere Canal.

The canal was closed in 1944 but the aqueduct is still in place.

24. TYRLEY CUTTING

HEW 463
SJ 694 316

Tyrley Cutting is a fine example of the scale of the earthworks on the Birmingham and Liverpool Junction Canal (HEW 1203, Chapter 6). It runs for about one and a quarter miles and is cut partly through rock. Presumably to reduce the quantity of excavation, the waterway is made narrower through the deep part of the cutting. Both at Tyrley cutting and at Woodseaves cutting about seven miles farther south there are characteristic arch bridges carrying roads as high as 70 ft over the canal.

CHAPTER 8

Cheshire and North Staffordshire

Cheshire is basically a shallow plain resembling a dish, which is split near the middle by a ridge of hills. The county is bounded on its eastern flank by the Pennines, on its western flank by the Clwydian Range, and the median sandstone ridge is crowned by the Delamere Forest. The northern geographical boundary is the River Mersey, though certain districts to the north of the river became part of the County after Local Government reorganization in 1974.

Cheshire has been an important corridor for trade since prehistoric times, and was of great military importance owing to its proximity to North Wales and the routes which led north to the Scottish border. Chester now stands where the Roman legionary fortress of Deva was built, and in mediaeval times the bridge over the River Dee was maintained by the Crown because of its strategic importance.

The industrial development of Cheshire and the county's salt and allied chemical industry was considerably aided by the construction of the canal network giving access to the sea routes, thus opening up new markets. From the 18th century the development of the turnpike road system, in which such eminent figures as Telford, Metcalfe and McAdam were involved, further assisted this process, and during the 19th century the spreading of the railway network again encouraged the expansion of industrial activity. Crewe grew into one of the largest railway towns in England, and Chester itself was the hub of six railway routes. In recent years the motorway construction programme has improved communications for the industrial areas and provided easier access to the scenery of North Wales for the development of tourism.

In the south-east of the region covered by this chapter lies the part of North Staffordshire dominated by the intensively industrialized area centred on Stoke-on-Trent and bounded on the east by the Pennine ridge. The county boundary between Cheshire and Staffordshire lies squarely on the watershed, the Cheshire

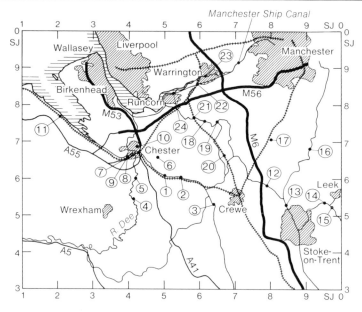

1. The Chester and Ellesmere Canals
2. Beeston Iron Lock
3. Nantwich Cast Iron Aqueduct
4. Holt Ancient Bridge, Farndon
5. Eaton Hall Iron Bridge
6. Hockenhull Packhorse Bridge
7. Chester Weir
8. Old Dee Bridge, Chester
9. Grosvenor Bridge, Chester
10. Chester Water Tower
11. Chester and Holyhead Railway
12. Trent and Mersey Canal
13. Harecastle Tunnels
14. Hazlehurst (Denford) Aqueduct
15. Cheddleton Mills
16. Macclesfield Canal
17. Jodrell Bank Radio Telescope
18. Weaver Navigation
19. Anderton Boat Lift
20. Grand Junction Railway
21. Dutton Viaduct
22. Acton Grange Viaduct
23. Warrington Bridge
24. Runcorn-Widnes Road Bridge

rivers flowing north to the Weaver and the Mersey, while the River Trent flows south through the Potteries. Within a relatively narrow corridor are found a variety of transport links, including the early Trent and Mersey (formerly Grand Trunk) Canal of 1777 with its pair of great tunnels at Harecastle, the Grand Junction Railway and, more recently, the M6 motorway.

1. THE CHESTER AND   HEW 1202
ELLESMERE CANALS   SJ 370 318 to SJ 405 774

These canals eventually formed part of the Shropshire Union Canal system. The $19\frac{1}{4}$ miles of the Chester Canal from the Dee at

Chester to Nantwich were built under the direction of Samuel Weston for boats 72 ft long and $13\frac{1}{2}$ ft wide, and were opened in 1779. The intention was to link with the Trent and Mersey Canal (HEW 1135) at Middlewich, but this was not achieved until 1833 with a branch from Wardle (SJ 613 571). The Chester Canal was therefore for some years in financial straits.

The Ellesmere Canal was authorized under an Act of 1793. Thomas Telford, then aged 36 and County Surveyor of Shropshire was, to use his own words, 'appointed Sole Agent, Architect and Engineer to the Canal',[1] with William Jessop as consulting engineer. It was proposed that the canal would link the Mersey, Dee and Severn via Ellesmere Port and Chester, passing through the North Wales coalfield and by way of Wrexham, Ruabon, Chirk and Ellesmere to Shrewsbury.

Work began on the $8\frac{3}{4}$ mile stretch from Chester to Netherpool (Ellesmere Port), known as the Wirral Line, which opened in 1797 and which merged with the Chester Canal in 1813 to become the Ellesmere and Chester Canal. Work began at the same time on the Llangollen Branch (HEW 1204, Chapter 1).

In 1800 a less expensive route from Chester was adopted, from the Chester Canal at Hurleston (SJ 626 553) north of Nantwich to Frankton (SJ 370 318), passing to the west of Whitchurch.

The canal port complex (HEW 1287, SJ 405 772) at Ellesmere Port, now leased to the Borough Council, is being vigorously rehabilitated by the Boat Museum Trust.

The Northgate Locks in the city of Chester (HEW 1337, SJ 402 666) are of interest because they are cut out of solid rock.

---

2. BEESTON IRON LOCK HEW 456
SJ 554 599

The Ellesmere and Chester Canals had, from the commencement of their construction, suffered from local failures caused by pockets of silt and running sand and in 1787 Beeston Lock collapsed because of these unstable conditions. Traffic ceased for a time because there were no funds available for repair.

In 1827 a short section of canal was re-aligned and two new locks were built, the lower one adjacent to the road (A49) being constructed entirely of flanged iron plates bolted together, as in Pontcysyllte Aqueduct (HEW 112, Chapter 1). This unique method of construction devised by Thomas Telford proved satisfactory, and no further stability problems were encountered.

3.   NANTWICH CAST IRON AQUEDUCT   HEW 497
SJ 642 526

Telford's aqueduct, constructed circa 1830, carries the Ellesmere
Canal over the Nantwich – Chester road with a headroom of 15 ft
6 in., and has a clear span of 29 ft 6 in. between masonry
abutments. Each side of the aqueduct proper is constructed of 6 ft
square flanged cast iron plates bolted together, providing a waterway of 13 ft with a 4 ft brick surfaced towpath. The whole is surmounted on both sides by cast iron railings 3 ft 9 in. high.

With the better known aqueduct at Stretton (HEW 228,
Chapter 6) over the A5 road, and a third at Congleton (HEW 494,
SJ 866 622) on the Macclesfield Canal, this work forms a trio all
designed by Telford to the same general pattern.

4.   HOLT ANCIENT BRIDGE, FARNDON   HEW 1221
Illustrated below   SJ 412 544

This bridge crosses the River Dee on the border between England
and Wales and between the villages of Farndon and Holt. It has
eight arches of about 24 ft span and a total length of 520 ft, which
includes the lengthy approach on the Welsh side. The width between parapets is only 13 ft, of which the carriageway takes up

Holt Ancient Bridge

10 ft. Seven of the arches are of normal two-ring construction, but the sixth arch from the English side has a further arch ring about 3 ft above the voussoirs, suggesting that there was a defensive tower over this span, which is known locally as 'The Lady's Arch', and also suggesting that the tower included a shrine dedicated to Our Lady. The date of this bridge is variously reported as 1345 and 1545.

5.   EATON HALL IRON BRIDGE   HEW 856
Illustrated on page 187   SJ 417 601

The private drive to Eaton Hall is carried across the Dee on a cast iron bridge constructed in 1824 a few yards upstream from the old ford (Aldford), by which the Roman Watling Street crossed the river. The single span of 150 ft consists of four arched ribs of I section, 36 in. deep in seven segments with ornate cruciform lattice bracing in the spandrels.

It was designed by Telford or based on his design for Craigellachie (HEW 24, NJ 286 452) and constructed by William Hazledine, the ironmaster, who worked with him on Pontcysyllte aqueduct and many other bridges. The excellent quality of the bridge is no doubt due to its aristocratic connections.

The 36 in. wide cast iron deck plates extend the full 17 ft width of the bridge and support some 14 in. of road metal.

The site is also an important point for gauging the flow of the River Dee.

6.   HOCKENHULL PACK HORSE BRIDGE   HEW 939
                                                           SJ 476 657

There are two packhorse bridges to the south-west of Chester – in Clwyd at Caergwrle and at Ffrith. Perhaps the one four miles east of Chester, at Hockenhull, is the more worthy of mention here, as it lies on the old route from Chester to London which ran rather to the south of the present A51 road. The river Gowy is crossed at Hockenhull by three separate arch spans of 17 ft in a total length of construction of some 240 ft.

It is shown on John Ogilvie's map of 1675 and remained in use until 1769 when it was by-passed. It is thought to be of 17th century construction, although there is reason to believe that there was a bridge at the site before 1353.

7. CHESTER WEIR   HEW 604
                                                          SJ 407 658

In 1071 the Norman Hugh d'Avranches (Hugh Lupus) became
Earl of Chester and it was he who built the first weir, on the eastern
side of the present Old Dee bridge, to provide the head of water
necessary to power the Dee mills. During early mediaeval times
the mills were destroyed by flood, but they were quickly rebuilt as
they belonged to the King and were the most important of the
many mills in the county. They even became famous nationally
and are remembered in the song 'The Jolly Miller of the Dee'.

The Dee mills were destroyed by fire in 1895 but the height of
the weir was raised early this century and used to power the hydro-
electric generating plant which was constructed on the site.

This plant no longer operates but the weir, which is approximately 170 yd long, still remains to give security to navigation
upstream and protect water supplies from salt and was until
recently the principal gauging point on the Dee.

8. OLD DEE BRIDGE, CHESTER   HEW 105
                                                          SJ 407 657

This red sandstone bridge has seven arched spans varying from
23 ft to 60 ft, one being semi-circular, two segmental and four
pointed. They surmount broad cutwaters, mostly with recesses
above. The present structure was built during the 14th century by
Henry de Snelleston, mason and surveyor to the Black Prince. It
has been rebuilt or repaired on many occasions since then. The last
major rebuilding was in 1826, when a footpath was added on its
eastern side. Until 1832 it was the only bridge across the River Dee
at Chester.

9. GROSVENOR BRIDGE, CHESTER   HEW 104
Illustrated on page 188   SJ 402 656

Thomas Harrison was the architect for Grosvenor Bridge about
the year 1802 but its construction was delayed for 25 years, by
which time he was nearly 80 years old. It is the crowning glory of
his career,[2] as, when it was formally opened by Princess, later
Queen, Victoria in 1832, it was the longest stone arch in the world,

Eaton Hall Iron Bridge

with a clear span of 200 ft.[3,4] It is still the longest in Britain and in the world is surpassed by only three others, all built in the 20th century, one at Plauen (295 ft) in East Germany, one at Salcano (279 ft) in Yugoslavia and the Pont Adolphe at Luxembourg (276 ft).[5] The arch was designed by George Rennie[6] and the whole bridge was built by James Trubshaw under the supervision of Jesse Hartley, using Peckforton stone, Scottish granite, and some Chester red stone. The bridge has a rise of 42 ft and carries a 24 ft wide carriageway between the parapets which are 33 ft apart.

The best view of the bridge is obtained from Edgar's Field on the south bank of the river, whence the Old Dee Bridge can also be seen.

---

10.   CHESTER WATER TOWER

Illustrated on page 190

HEW 1080

SJ 416 666

Since the incorporation of the City of Chester Waterworks in 1826, water has been taken from the River Dee at the Barrel Well site and until 1853 it was pumped untreated into the supply system.

The present Tarvin Road Waterworks was then built to the designs of J. F. La Trobe Bateman. The works included a reservoir, three slow sand filters, a pure water tank, a pumphouse and a water tower. The last two and one of the sand filters still exist.

The 70 ft dia. brick tower was originally 64 ft high, surmounted by a 12 ft deep cast iron tank with a capacity of 268 000 gallons. In 1889, a further 20 ft head of water to meet the demands of the expanding city was achieved by jacking up the tank and inserting additional brickwork between the original top of the tower and the base of the tank.

| 11. | **CHESTER AND** | **HEW 1094** |
| | **HOLYHEAD RAILWAY** | SJ 413 760 to SH 248 822 |

The Chester and Holyhead Railway, already dealt with in some detail in Chapter 1, began at Chester General Station. The imposing range of station buildings by Francis Thompson, though altered several times, still remains.

After leaving the station, the line passed through a tunnel which was later opened out to form a cutting and two more tunnels, that under Windmill Hill being unusual in that it accommodates four tracks. (Richmond Hill Tunnel at Leeds (HEW 234, SE 314 334) has five.)[7]

The Chester and Holyhead Railway then crosses the River Dee on a three-span girder bridge, originally cast iron with wrought iron bracing. On 24 May 1847 one of the 98 ft spans collapsed under load. The inquiry which followed did much to hasten the

Grosvenor Bridge

development of straightforward wrought iron and, later, steel plate girders.

---

12. TRENT AND MERSEY CANAL

HEW 1135
SJ 857 500 to SJ 567 810

The Trent and Mersey Canal, which has been dealt with more extensively in Chapter 6, passes under the high ground between Kidsgrove and Tunstall, north-east of Stoke-on-Trent, by means of the Harecastle Tunnels before beginning the descent via Middlewich and Northwich with 36 locks (originally 35) to the Bridgewater Canal (HEW 976) and Preston Brook (SJ 567 810).

There are also tunnels at Barnton, 572 yd long (SJ 634 748), Saltersford, 424 yd long (SJ 626 753) and Preston Brook, 1239 yd long (SJ 572 794).

The Caldon Canal from the Trent and Mersey at Stoke-on-Trent to Froghall (SK 022 475) was opened in 1778.

---

13. HARECASTLE TUNNELS

Illustrated on page 193

HEW 54, HEW 465
SJ 849 517 to SJ 837 541

The original Harecastle Tunnel (HEW 54) was engineered by one of the pioneer canal engineers, James Brindley, and its 1 mile 1120 yd length took nine years to construct between 1766 and

1775. It was the first tunnel in Great Britain to be constructed solely for transport purposes and even now ranks among the longest canal tunnels built. Fifteen working shafts were used in its construction, enabling work to be carried out at 32 different faces. Many problems were encountered during the driving of the tunnel, not the least of which was the ingress of large quantities of water. Eventually, steam pumps were used to clear the workings.

This tunnel was only 8 ft 6 in. wide and did not have a towpath, so that boats had to be laboriously 'legged' through. Owing to the restricted width, boats could not pass simultaneously in both directions, which caused a severe bottleneck on the canal as traffic built up. Eventually a decision was made to build a new tunnel alongside the old one.

Harecastle New Tunnel (HEW 465), 1 mile 1166 yd long, is slightly longer than its predecessor and lies to the east thereof. Engineered by Thomas Telford, it was built in two years and opened in 1827.

The New Tunnel is wider and has a towpath. Both tunnels were used for a time, carrying traffic in opposite directions but the Old Tunnel was closed in 1918 and is now derelict owing to damage caused primarily by mining subsidence. The New Tunnel remains

Chester Water Tower

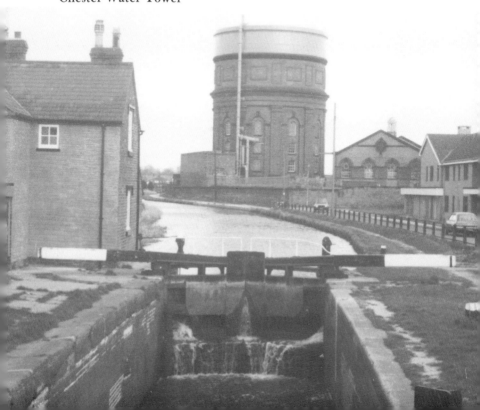

in use, although the size of boats which may pass through is limited by subsidence within the tunnel, and the towpath is now impassable.

| **14. HAZLEHURST (DENFORD)** | **HEW 554** |
| **AQUEDUCT** | **SJ 954 536** |

Flyovers are now commonplace on motorways and have been in use on railways for many decades. They are also to be found on canals as witness the Hazlehurst 'Flyover' on the Leek branch of the Caldon Canal in Staffordshire.

This aqueduct was necessary because of a realignment in the local canal system on the opening of the Stoke to Leek railway in 1841.

The Leek canal branch diverges slightly to the south and then for 660 yd runs parallel to the parent canal, which over this distance has been lowered about 26 ft by a flight of three locks. At this point the branch bends to the north-east, and crosses over the main canal and the railway by means of an attractive brick arch aqueduct.

Another attractive feature of this site is the cast iron arch roving bridge (HEW 857) at SJ 947 538 between the junction and the locks.

| **15. CHEDDLETON MILLS** | **HEW 1133** |
| | **SJ 973 526** |

On the banks of the River Churnet at Cheddleton in Staffordshire is an interesting pair of water-driven flint-grinding mills. Ground flint is used extensively in the pottery industry.

When the mills were built is not known, but there are records of a corn mill on the site from as early as the 13th century. The older, the South mill, was at some time converted from corn- to flint-grinding. The North mill was built specifically for the flint trade and may have been the work of James Brindley. There is a mill by Brindley dated 1752 at Leek (SJ 977 569).

The 'low-breast' wheels are 20 ft 5 in. (South mill) and 22 ft (North mill) in diameter. They are built from cast-iron segmental rims with oak spokes connecting the rims to cast iron hubs which are mounted on cast iron shafts. The shaft of the North mill is cruciform in section, $15\frac{1}{4}$ in. in size, whilst the South mill has a hexagonal shaft 16 in. across the flats. The motion of each wheel is transmitted to the vertical driving shaft by two great bevelled gear wheels, one 10 ft dia. and one 8 ft 6 in. dia. in the North mill, and both 6 ft 6 in. dia. in the South mill. In addition to driving the grinding pan, the power of the wheel is also used to drive a hoist mechanism and pumps.

The mills have both been restored and are now maintained in working order by a Trust. They may be viewed at weekends.

---

16. MACCLESFIELD CANAL   HEW 1136
SJ 836 546 to SJ 961 884

The Macclesfield Canal was born too late, right at the end of the Canal Era. Its route was surveyed by Thomas Telford early in 1825, and the enabling Act received Royal Assent in April 1827. It was completed in 1831, at the beginning of the Railway Era, and like the Birmingham and Liverpool Junction Canal (HEW 1203, Chapter 6) was designed not to meander along the contours of the land but to take a straighter course on embankments. Nevertheless, it was never able to take full advantage of its 'modern' design, partially because of the jealousies of existing canal companies, particularly the Trent and Mersey (HEW 1135). The Engineer, Thomas Brown, like Telford, appears to have played little or no part in its construction, the works being supervised by William Crosley the Younger.

The canal provided a link from the Potteries in the south to Manchester in the north, joining the Trent and Mersey Canal at a junction at Hall Green near Kidsgrove, and the Peak Forest Canal above the top lock at Marple. Roughly following the 400 ft contour, the many embankments involved heavy earthworks. The rise was concentrated in one flight of twelve locks at Bosley (HEW 1314, SJ 906 663). The stop lock at Hall Green is a relic of the jealousies between the various canal companies. This adjusts the level of the water by only 1 ft to prevent any of the Trent and Mersey's summit water from entering the Macclesfield Canal.

---

17. JODRELL BANK RADIO TELESCOPE   HEW 471
Illustrated on page 194   SJ 795 711

The University of Manchester's radio telescope at Jodrell Bank is a familiar feature of the Cheshire landscape. The need for a fully-

Harecastle Tunnels: north entrances, with old tunnel on right and new tunnel on left

steerable instrument of such a size developed as a result of the limitations of the 218 ft transit telescope built in 1947 which was its predecessor.

The telescope was designed by Husband and Company of Sheffield to meet the exacting requirements of the astronomer Professor Bernard Lovell.[8]

Thirty different firms were engaged on the construction, which was completed in 1957. A paraboloid bowl 250 ft in dia. formed of welded steel sheets is carried on a space frame of structural steel. Radio signals from space are reflected to the focus of the bowl where aerials are mounted on a steel mast 62 ft 6 in. high. The bowl and its supporting structure are carried by trunnion bearings which were formerly part of the 15 in. gun turrets of HMS *Revenge* and HMS *Royal Sovereign*.

The two supporting towers which carry the bowl assembly at an elevation of 165 ft are braced to form a yoke under the bowl, and are mounted on bogies on a track of two concentric rails of overall diameter 353 ft. The telescope can thus be rotated to point at any portion of the sky.

The moving parts weigh approximately 2000 tons and the whole structure nearly 3200 tons, following modifications made in 1970. The loads on the trunnion bearings and the rail tracks were then reduced by the incorporation of two semicircular wheel girders under the bowl that transferred one third of the load to a new central rail track.

There is a museum and a planetarium on the site, but the telescopes and their control rooms are not open to the general public.

---

18.   WEAVER NAVIGATION   HEW 1191
SJ 655 664 to SJ 493 818

The River Weaver flows northwards from central Cheshire through Nantwich, Winsford and Northwich to join the River Mersey at Weston Point near Runcorn.

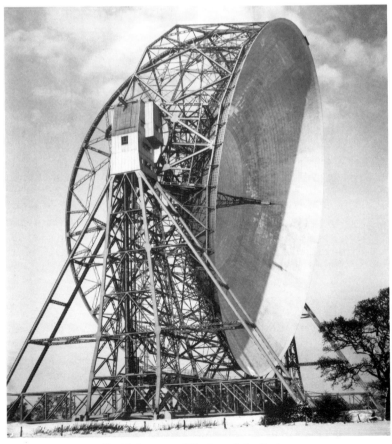

Jodrell Bank Radio Telescope (Husband & Co.)

The last seven miles of the river were always navigable, and substantial improvements in the 1730s enabled 40-ton craft to reach Winsford 20 miles upstream, the Navigation passing through eleven locks en route. The main traffic was coal and salt. Problems with the tidal entrance to the Mersey were overcome by Thomas Telford in 1810, when the four-mile Weston Canal and a new basin and river lock were opened at Weston Point. Improvements recommended by William Cubitt were made in the 1840s to enable the Navigation to be used by 100-ton flats. Enlarged locks designed by E. L. Williams, Junior enabled coastal traffic to use the Navigation in 1870.

At the end of the 19th century a new dock was built at Weston Point and a number of the smaller locks were replaced by fewer and larger (229 ft by 42 ft 6 in.) locks with gates operated by Pelton wheels. The construction in 1875 of a boat lift at Anderton (HEW 286) enabled interchange of traffic with the Trent and Mersey Canal (HEW 1135) and in 1894 the opening of the Manchester Ship Canal (HEW 88)[9] gave a new outlet to deeper water at Eastham.

Interesting structures associated with the Weaver Navigation in addition to those already mentioned are Dutton Horse Bridge (HEW 1316, SJ 584 767), the railway viaducts at Frodsham (HEW 1168, SJ 529 786 and SJ 534 790), and several swing bridges including Acton Bridge (HEW 1281, SJ 601 761), Sutton Weaver Bridge (SJ 535 788) and Town (SJ 656 738) and Hayhurst (Navigation) (SJ 656 736) at Northwich (HEW 1282). These swing bridges, designed by J. A. Saner, have the unusual feature of being supported partly on piles and partly on floating caissons, to counteract subsidence from salt mine workings.[10]

Commercial traffic to Winsford ceased in the 1950s, but coasters up to 800 tons still navigate as far as Northwich.

---

19. ANDERTON BOAT LIFT   **HEW 286**
Illustrated on page 196   SJ 647 752

This lift, which is unique in Britain, transfers canal boats between the River Weaver Navigation and the Trent and Mersey Canal through a 50 ft 4 in. difference in water levels.[11,12]

At the top level the lift is linked to the canal by means of a 162 ft long aqueduct which has twin channels.

The lift was originally operated by water pressure and the two caissons were supported on 3 ft dia. rams working in the cylinders of the hydraulic presses which were linked through valves. The caissons are of wrought iron, being 75 ft long and 15 ft 6 in. wide, with gates at each end. By 1904 the lift, which was opened in 1875, needed overhauling and the main rams required renewing. It was therefore converted to a counter-balanced structure driven by electric motors, with the caissons suspended from wire ropes carrying cast iron weights, some 252 tons to each caisson.[13]

The lift, which is constructed of wrought and cast iron, was designed by Sir Edward Leader Williams, Edwin Clark and J. W. Sandeman, and was the forerunner of several similar lifts constructed in Europe.

The Anderton lift, although still in use, is now a scheduled Ancient Monument.

---

20. GRAND JUNCTION   HEW 1129
**   RAILWAY   SJ 728 500 to SJ 578 951**

The Grand Junction Railway, introduced in Chapter 6 as Item 13, was the fruit of the experience of George Stephenson and the young men about him in perfecting their craft of railway construction, learnt in building the Liverpool and Manchester Railway (HEW 223)[14] and in maintaining and improving that pioneering work. On the northern part of the line, perhaps the most important features are the junction at Crewe, with its station famed in a

Anderton Boat Lift (British Waterways Board)

music-hall song, and the adjacent Locomotive Works (HEW 276, SJ 709 552 to SJ 686 561) begun in 1842 in the angle between the main line and the Chester and Crewe Railway, which was absorbed by the Grand Junction Railway during its construction.

The success of the steam locomotive in the 1830s gave the railways a virtual monopoly of inland transport from 1840 until met by competition from motor car and lorry around 1910. The widespread network of main lines, cross-country routes and branch lines demanded the support of vast engineering organizations met by the setting up of Main Works of which Crewe and Swindon (Chapter 4, Item 12) were major examples. In their heyday in the 1880s, Crewe Works extended for about two miles, although to-day little remains on site of historical interest.[15]

Like other works, Crewe designed, produced and maintained not only locomotives, carriages and wagons but prefabricated buildings, bridges, rails, points and crossings; provided chemical services; even had its own gas works, and for a time from 1864 produced steel. The works were self-supporting in housing for the staff and provided professional training in mechanical engineering of world wide importance.

The first locomotive, 'Columbine', produced at Crewe is preserved in the National Railway Museum at York. Among the mechanical engineers in charge of the Crewe works may be mentioned Francis Trevithick (1847 – 57), son of Richard Trevithick of Penydarren Tramroad fame (Chapter 3, Item 14); John Ramsbottom (1857 – 71) who installed the first ever pick-up water troughs (Chapter 1, Item 8); and F. W. Webb (1871 – 1903), who was involved in the standardization not only of station buildings (Chapter 1, Item 10) but also of small cabins and office blocks.

North of Crewe, the Grand Junction Railway Company built some notable viaducts such as those at Vale Royal (SJ 643 706), with five arches of 62 ft span, Dutton (HEW 154) and the twelve-arch structure over the Mersey south of Warrington at SJ 600 866. This last bridge ceased to serve the main line when the Manchester Ship Canal was built and the railway was diverted to a new high level bridge at Acton Grange (HEW 1229).

---

21. DUTTON VIADUCT   HEW 154
Illustrated on page 198   SJ 582 764

Dutton Viaduct was designed by Joseph Locke to carry the Grand Junction Railway across the valley of the River Weaver. It was built between 1834 and 1836 at a cost of £54 440 and opened in 1837. The contractor was David McIntosh.

It is of masonry construction and consists of twenty segmental arches of 60 ft span with a rise of 17 ft 6 in. Rail level is at a maximum height of 65 ft. Today it carries the two electrified tracks of British Rail's West Coast Main Line.

---

22.   ACTON GRANGE VIADUCT   HEW 1229
Illustrated on facing page   SJ 590 857

The Manchester Ship Canal (HEW 88) currently has twenty bridges across it, from the big railway viaduct at Runcorn in the west (HEW 196, SJ 509 835)[16] to an insignificant small bridge at the Manchester end. Nine are swing bridges; eleven are fixed; fifteen have the original (1893) superstructures.

Apart from the Barton Aqueduct (HEW 28, SJ 767 976)[17] perhaps the most notable is the high level fixed span carrying the West Coast main line of British Rail. An opening span would not have been suitable so the line of the then London and North Western Railway, originally the Grand Junction Railway (HEW 1129), was diverted to the west over a considerable distance, to

Dutton Viaduct and Dutton Horse Bridge (Chris Edge)

Acton Grange Viaduct (British Rail)

enable it to climb at an acceptable gradient to clear the Ship Canal above the mast height of ships.

A branch line to Chester diverges at this point and this also had to be carried on the new viaduct which is on a very considerable skew; the skew span, between masonry abutments, being 263 ft to provide a square span of only 120 ft.

The type of structure chosen was a form of lattice girder known as a Whipple-Murphy truss, very popular for a period around 1890 – 1910. There are five main girders.

---

23. WARRINGTON BRIDGE

HEW 254
SJ 608 879

This bridge, which carries the A49 road, was built in 1915 and is the sixth bridge to occupy the same site at this important crossing of the River Mersey, which is still tidal at this point, where a bridge was recorded as early as 1305.

The bridge has a clear span of 134 ft and is 80 ft wide between the parapets. Eight reinforced concrete parabolic arch ribs carry the deck, with a shallow rise to span ratio of 1 to 10, dictated by the level of the road. The lateral thrusts of 350 tons at the abutments are taken by reinforced concrete counterforts at each rib, bearing on inclined reinforced concrete piles driven into the hard clay on the line of the resultant thrust.

Temporary concrete hinges were used to accommodate 3 in. of settlement that took place during the construction period. This was perhaps the first bridge to be built in Britain with reinforced concrete hinges, which were located at the crowns of the arch ribs and reinforced by shear bars and interlocking spirals of reinforcement. The thrust on these hinges was about 240 tons.

The deck loading is transferred to the ribs by means of short columns, and the bridge was subjected to carefully monitored test loadings using a combination of tramcars and steam rollers before it was opened. At high water the springings of the arch ribs are under water, which gives some idea of the physical constraints of the site which faced the designers, Webster and Fitzsimons, who acted in conjunction with Considère Constructions Ltd. In addition, so as not to interfere with road traffic, the bridge had to be built in two halves. It was constructed by Alfred Thorne and Sons.

Runcorn-Widnes Road Bridge (Cheshire County Council)

24. RUNCORN – WIDNES ROAD BRIDGE   HEW 1063
Illustrated on facing page   SJ 510 835

The main span of 1082 ft of this elegant modern bridge, which
crosses the River Mersey with a clearance of 75 ft, is longer than
that of any other steel arch in Britain and is twice the span of the
Tyne Bridge at Newcastle (HEW 91, NZ 254 638)[18] which was the
longest at the time of its opening in 1928.

Mott, Hay and Anderson designed the bridge to replace the
former transporter bridge which was built in 1905 and was the
longest (1000 ft span) and the lowest (75 ft) of the five built in this
country.[19] A two-pinned bowstring arch design was adopted after
wind tunnel model tests indicated that the presence of the adjacent
railway bridge (HEW 196) could, under certain wind conditions,
produce severe oscillations if a conventional suspension bridge
design were used.

Leonard Fairclough Ltd and Dorman Long Ltd commenced
construction in April 1956 and the bridge was opened in July 1961.

The bridge has side spans of 250 ft, one crossing the Manchester
Ship Canal with a clearance of 80 ft. It was widened between 1975
and 1977.

Bibliography and references

General bibliography

Abbott W. *The turnpike road system in England and Wales 1663–1840.* Cambridge University Press, 1972.

Adamson S.H. *Seaside piers.* Batsford, London, 1977.

Barbey M.F. *Civil engineering heritage: northern England.* Thomas Telford, London, 1981.

Barrie D.S.M. *A regional history of the railways of Great Britain.* Vol. 12: South Wales. David and Charles, Newton Abbot, 1980.

Baughan P.E. *The Chester and Holyhead Railway.* David and Charles, Newton Abbot, 1972.

Baughan P.E. *A regional history of the railways of Great Britain.* Vol. 11: North and Mid-Wales. David and Charles, Newton Abbot, 1980.

Beaver P. *A history of lighthouses.* Peter Davies, London, 1971.

Beckett D. *Brunel's Britain.* David and Charles, Newton Abbot, 1980.

Beckett D. *Stephenson's Britain.* David and Charles, Newton Abbot, 1984.

Berridge P.S.A. *The girder bridge.* Robert Maxwell, London, 1969.

Binnie G.M. *Early Victorian water engineers.* Thomas Telford, London, 1981.

Blower A. *British railway tunnels.* Ian Allan, London, 1964.

Bourne J.C. *The history and description of the Great Western Railway.* Bogue, London, 1864; reissued David and Charles, Newton Abbot, 1970.

Bracegirdle B. *The archaeology of the Industrial Revolution.* Heinemann, London, 1973.

Brees S.C. *Railway practice,* 2nd series. John Williams, London, 1846.

———. *British bridges.* Public Works, Roads and Transport Congress, London, 1933.

Brunel I. *The life of Isambard Kingdom Brunel, civil engineer.* Longman, London, 1870; reissued David and Charles, Newton Abbot, 1971.

Christiansen R. *A regional history of the railways of Great Britain.* Vol. 7: the West Midlands. David and Charles, Newton Abbot, 1973.

Christiansen R. *A regional history of the railways of Great Britain.* Vol. 13: Thames and Severn. David and Charles, Newton Abbot, 1981.

Conder F.R. *Personal recollections of English engineers.* Hodder and Stoughton, London, 1868. Reissued as *The men who built railways.* Thomas Telford, London, 1983.

Cossons N. *The BP book of industrial archaeology*. David and Charles, Newton Abbot, 1975.

Cossons N. and Trinder B. *The Iron Bridge*. Moonraker Press, Bradford-on-Avon, 1979.

Cresy E. *Encyclopaedia of civil engineering*. Longman, London, 1856.

de Maré E. *Bridges of Britain*. Batsford, London, 1975.

Dempsey G.D. *Tubular and other iron bridges*. Virtue Bros., 1864.

Devereux R. *John Loudon McAdam: chapters in the history of highways*. Oxford University Press, London, 1936.

Dunn J.M. *The Chester and Holyhead Railway*. Oakwood Press, Trowbridge, 1968.

Fairbairn W. *The application of cast iron and wrought iron to building purposes*. Weale, London, 1857.

Fowler C.E. *The ideals of engineering architecture*. Spon, London, 1929.

Gladwin D.D. and Gladwin J.M. *Canals of the Welsh valleys, and their tram roads*. Oakwood Press, Blandford, 1974.

Hadfield C. *The canals of South Wales and the Border*. David and Charles, Newton Abbot, 1967.

Hadfield C. *British canals: an illustrated history*. David and Charles, Newton Abbot, 1979.

Hadfield C. *The canal age*. David and Charles, Newton Abbot, 1981.

Hadfield C. and Skempton A.W. *William Jessop, engineer*. David and Charles, Newton Abbot, 1979.

Hague D.B. and Christie R. *Lighthouses, their architecture, history and archaeology*. Gomer Press, Llandyssul, 1975.

Harris R. *Canals and their architecture*. Godfrey Cave, London, 1980.

Heyman J. *The masonry arch*. Ellis Horwood, Chichester, 1982.

Humber W. *Cast iron and wrought iron bridges and girders for railway structures*. Spon, London, 1857.

Jackson D. *Lighthouses of England and Wales*. David and Charles, Newton Abbot, 1978.

Jensen M. *Civil engineering around 1700*. Danish Technical Press, Copenhagen, 1969.

Jervoise E. *Ancient bridges of Wales and western England*. Architectural Press, London, 1936.

Jones E. *The Penguin guide to the railways of Britain*. Allen Lane, London, 1981.

Latimer J. *Annals of Bristol in the 19th century*, vol. 3. Kingsmead reprints, Bath, 1970.

Lindsay J. *The Trent and Mersey Canal*. David and Charles, Newton Abbot, 1977.

MacDermot E.T. and Clinker C.R. *History of the Great Western Railway*. Vol. 1: 1833–63; vol. 2: 1863–1921. Ian Allan, London, 1964.

Maggs C. *Rail centres, Bristol*. Ian Allan, London, 1981.

Margary J.D. *Roman roads in Britain*. Phoenix House, London, 1957.

Marshall J. *The Guinness book of rail facts and feats*. Guinness Superlatives, Enfield, 1979.

Nock O.S. *The railway engineers.* Batsford, London, 1955.

Paar H.W. *A history of the railways of the Forest of Dean.* Vol. 1: Severn and Wye. David and Charles, Newton Abbot, 1971.

Paar H.W. *A history of the railways of the Forest of Dean.* Vol. 2: Great Western. David and Charles, Newton Abbot, 1975.

Penfold A. (ed.). *Thomas Telford: engineer.* Thomas Telford, London, 1980.

Prideaux J.D.C.A. *The Welsh narrow gauge railway: a pictorial history.* David and Charles, Newton Abbot, 1976.

Priestley J. *Navigable rivers, canals and railways throughout Great Britain.* Longman, London, 1831.

Pugsley Sir A. (ed.). *The works of Isambard Kingdom Brunel: an engineering appreciation.* Institution of Civil Engineers and University of Bristol, London and Bristol, 1976.

Reader W.J. *Macadam: The McAdam family and the turnpike roads 1798 – 1861.* Heinemann, London, 1980.

Rennie Sir J. *British and foreign harbours.* Weale, London, 1854.

Rolt L.T.C. *Isambard Kingdom Brunel.* Pelican, London, 1972.

Rolt L.T.C. *George and Robert Stephenson.* Pelican, London, 1978.

Smith N. *A history of dams.* Peter Davies, London, 1971.

Stephens J.H. *The Guinness book of structures.* Guinness Superlatives, Enfield, 1976.

Stevens R.A. *Towpath guide to the Brecknock and Abergavenny and Monmouthshire canals.* Goose and Son, Cambridge, 1974.

Telford T. (Rickman J. (ed.)). *Life of Thomas Telford . . . with a folio atlas of copper plates.* London, 1838.

Walters D. *British railway bridges.* Ian Allan, London, 1963.

Walmisley A.T. *Iron roofs.* Spon, London, 1888.

Webb S. and Webb B. *English local government: the story of the King's highway.* Longmans, 1913; reissued Frank Cass, London, 1963.

Wishaw F. *The railways of Great Britain and Ireland.* Simpkin Marshall, London, 1840; reissued David and Charles, Newton Abbot, 1969.

Wright G.H. *Bridges of Britain: a pictorial survey.* D. Bradford Barton, Truro, 1973.

Wryde J.S. *British lighthouses.* Unwin, London, 1913.

Chapter 1: North Wales

1. Hawkshaw Sir J. *Holyhead new harbour.* Final report to the Board of Trade. 1873.

2. Hayter H. Holyhead new harbour. *Min. Proc. Instn Civ. Engrs,* 1875 – 76, Part 2, **44**, 95 – 130.

3. Telford T. Reports to the Commissioners for the road from London to Holyhead, 1824 – 34.

4. Paxton R.A. Menai Bridge (1818 – 1826) and its influence on suspension bridge development. *Trans. Newcomen Soc.,* 1977 – 78, **49**, 87 – 110.

5. Maude T.J. Account of the alterations made on the structure of the

Menai Bridge, during the repairs in consequence of the damage it received from the gale of January 7, 1839. *Trans. Instn Civ. Engrs*, 1841, **3**, Part 5, 371–375.

6. Provis W.A. Observations on the effects of wind on the suspension bridge over the Menai Strait, more especially as relates to the injuries sustained by the roadways during the storm of January, 1839. *Trans. Instn Civ. Engrs*, 1841, **3**, Part 5, 357–370.

7. Maunsell G.A. Menai Bridge reconstruction. *J. Instn Civ. Engrs*, 1946, **25**, Jan., No. 3, 165–206.

8. Husband J. The aesthetic treatment of bridge structures. *Min. Proc. Instn Civ. Engrs*, 1901, **145**, 166.

9. Provis W.A. *An historical and descriptive account of the suspension bridge constructed over the Menai Strait in North Wales*. Ibbotson and Palmer, London, 1828.

10. Provis W.A. Observations on the effects of wind on the suspension bridge over the Menai Strait, more especially as relates to the injuries sustained by the roadways during the storm of January, 1839. *Trans. Instn Civ. Engrs*, 1841, **3**, 368.

11. Ferguson H. The 'preferred' route nobody prefers. *New Civ. Engr*, 1975, 17 July, 28–29.

12. Barbey M.F. *Civil engineering heritage: northern England*. Thomas Telford, London, 1981, 137–143.

13. Parry E. *Railway companion from Chester to Holyhead*. Catherall, Chester, 1849, 2nd edn; facsimile by E. & W. Books, London, 1970.

14. Fox F. The Hawarden bridge. *Min. Proc. Instn Civ. Engrs*, 1892, **108**, Part 2, 304–317.

15. Hawarden Dee railway bridge. *Ill. Lond. News*, 1889, 10 Aug., 184.

16. Parkinson J. A55 coast road raft floats on sea of piles. *New Civ. Engr*, 1983, 27 Jan., 20.

17. Greeman A. Coast road treads softly through Colwyn Bay. *New Civ. Engr*, 1984, 9 Feb., 30–32 and cover.

18. Clark E. *Britannia and Conway tubular bridges*. Day, London, 1850.

19. Fairbairn W. Account of the construction of the Britannia and Conway tubular bridges. *Min. Proc. Instn Civ. Engrs*, 1850, **9**, 233–287.

20. Seyrig T. The different modes of erecting iron bridges. *Min. Proc. Instn Civ. Engrs*, 1880–81, Part 1, **63**, 161–162.

21. Fowler C.E. *The ideals of engineering architecture*. Spon, London, 1929, 171.

22. Husband H.C. Reconstruction of the Britannia Bridge. *Proc. Instn Civ. Engrs*, Part 1, 1975, **58**, Feb., 25–66.

23. Hobson G.A. The Victoria Falls Bridge. *Min. Proc. Instn Civ. Engrs*, 1907, Part 4, **170**, 1–23.

24. Barbey M.F. *Civil engineering heritage: northern England*. Thomas Telford, London, 1981, 118.

25. Simmons J. *Transport museums in Britain and Western Europe*. Allen & Unwin, London, 1970, 141.

26. Baines J.A. *et al.* Dinorwig pumped storage scheme. *Proc. Instn Civ. Engrs*, Part 1, 1983, **74**, Nov., 635–718.
27. Smith W. Summit-level tunnel of the Bettws and Festiniog Railway. *Min. Proc. Instn Civ. Engrs*, 1882–83, Part 3, **73**, 150–177.
28. Dunn J.M. From Llandudno Junction to Blaenau Festiniog. *Rly Mag.*, 1959, **105**, Dec., 821–822.
29. Boyd J.C.L. *The Festiniog Railway*. Odhams Press, London, 1962.
30. Lewis M.S.T. *How Festiniog got its railway*. Railway and Canal Historical Society, Caterham, 1965.
31. Thomas Telford Ltd. Funds galvanised Festiniog. *New Civ. Engr*, 1984, 9 Feb., 36–37.
32. Roseveare J.C.A. Festiniog pumped storage scheme. *Proc. Instn Civ. Engrs*, 1964, **28**, May, 1–30.
33. Conybeare H. Description of viaducts across the estuaries on the line of the Cambrian Railway. *Min. Proc. Instn Civ. Engrs*, 1870–71, Part 2, **32**, 137–145.
34. Thomas Telford Ltd. In brief. *New Civ. Engr*, 1982, 4 Nov., 7.
35. Deacon G.F. The Vyrnwy works for the water-supply of Liverpool. *Min. Proc. Instn Civ. Engrs*, 1895–96, Part 4, **126**, 24–67.
36. Deacon G.F. The Vyrnwy works and the Vyrnwy valley water supply to Liverpool. *Engineer, Lond.*, 1892, **73**, 15 July, supplement, 13–15.
37. Stilgoe J.H.T. Trunk mains for water supply. *Trans. Lpool Engng Soc.*, 1949, **72**, Feb., 117–151.
38. White W.F. Vyrnwy aqueduct, fourth instalment pipeline. *J. Instn Wat. Engrs*, 1950, **4**, Feb., 13–67.
39. Hadfield C. and Skempton A.W. *William Jessop, Engineer*. David and Charles, Newton Abbot, 1979, 222–228.

Chapter 2: Mid-Wales
1. Mansergh E.L. and Mansergh W.L. The works for the supply of water for the city of Birmingham from Mid-Wales. *Min. Proc. Instn Civ. Engrs*, 1911–12, Part 4, **190**, 23–25.
2. Carlyle W.J. and Owen R.C. The River Towy scheme for the West Glamorgan Water Board. *J. Instn Wat. Engrs*, 1973, **27**, Aug., 287–317.

Chapter 3: South Wales
1. Barbey M.F. *Civil engineering heritage: northern England*. Thomas Telford, London, 1981, 7.
2. Tucker D.G. Half a century of hydro-electricity at Monmouth. *J. Monmouthshire Local History Council*, 1974, **37**, Spring, 27–37.
3. Harris P.G. *Wye valley industrial history*. Privately printed, 1976, 33–38.
4. *Diary of Thomas Jenkins of Llandeilo 1826–1870*. Dragon Books, Bala, 1976.
5. Binnie G.M. *Early Victorian water engineers*. Thomas Telford, London, 1981, 157–201.

6. Fletcher L.E. Description of the Landore viaduct on the line of the South Wales Railway. *Min. Proc. Instn Civ. Engrs*, 1854–55, **14**, 492–506.

7. Whitley H.S.B. The reconstruction of Carmarthen Bridge. *Min. Proc. Instn Civ. Engrs, 1916–17*, **204**, 365–368.

8. Barbey M.F. *Civil engineering heritage: northern England.* Thomas Telford, London, 1981, 132.

9. Barbey M.F. *Civil engineering heritage: northern England.* Thomas Telford, London, 1981, 40–41.

10. Atkins W.S. Transportation – road and rail. *Civil engineering problems of the South Wales valleys.* Institution of Civil Engineers, London, 1970, 41–54.

11. Smith T.M. Account of the Pont-y-tu-prydd over the River Tâfe near Newbridge in the county of Glamorgan. *Min. Proc. Instn Civ. Engrs*, 1846, **5**, 474–477.

12. Bruce G.B. The Royal Border Bridge. *Min. Proc. Instn Civ. Engrs*, 1850–51, **10**, 241–242.

13. Bell W. On the stresses of rigid arches, continuous beams, and curved structures. *Min. Proc. Instn Civ. Engrs*, 1871–72, Part 1, **33**, 58–63.

14. Hague D. and Hughes S. Pont y Cafnau, the first iron railway bridge and aqueduct? *Ass. Industrial Archaeology Bull.*, 1982, **9**, No. 4, 3–4.

15. The Mumbles Railway Society. *The Mumbles Railway, the world's first passenger railway.* R. Cottle, Swansea, 1981.

16. Lee C.E. *The first passenger railway – the Oystermouth or Swansea and Mumbles Line.* Railway Publishing Co., London, 1942.

17. Robinson J. The Barry Dock works, including the hydraulic machinery and the mode of tipping coal. *Min. Proc. Instn Civ. Engrs*, 1889–90, Part 3, **101**, 129–151.

18. Robinson J. The Barry Graving Docks. *Min. Proc. Instn Civ. Engrs*, 1893–94, Part 2, **116**, 267–274.

19. Maynard H.N. *Handbook to the Crumlin Viaduct.* Wilson, Crumlin, 1862.

20. Husband J. The aesthetic treatment of bridge structures. *Min. Proc. Instn Civ. Engrs*, 1900–01, Part 3, **145**, 217.

21. Cubitt J. A description of the Newark Dyke Bridge on the Great Northern Railway. *Min. Proc. Instn Civ. Engrs*, 1852–53, **12**, 601–612.

22. Maynard H.N. Discussion: The different modes of erecting iron bridges, by T. Seyrig. *Min. Proc. Instn Civ. Engrs*, 1880–81, Part 1, **63**, 189–191.

23. The transporter bridge at Newport. *Engineer, Lond.*, 1906, **102**, 14 Sept., 263–265. (Abstract: *Min. Proc. Instn Civ. Engrs*, 1906–07, Part 1, **167**, 405–406.)

24. Barbey M.F. *Civil engineering heritage: northern England.* Thomas Telford, London, 1981, 48–50.

25. Barbey M.F. *Civil engineering heritage: northern England.* Thomas Telford, London, 1981, 161–162.

Chapter 4: Avon and North Wiltshire

1. Roberts G. Severn Bridge: design and contract arrangements. *Proc. Instn Civ. Engrs*, 1968, **41**, Sept., 1–48.
2. Gowring G. and Hardie A. Severn Bridge: foundations and substructure. *Proc. Instn Civ. Engrs*, 1968, **41**, Sept., 49–67.
3. Hyatt K.E. Severn Bridge: fabrication and erection. *Proc. Instn Civ. Engrs*, 1968, **41**, Sept., 69–104.
4. Barbey M.F. *Civil engineering heritage: northern England.* Thomas Telford, London, 1981, 14–16.
5. Anderson J.K. The Tamar Bridge. *Proc. Instn Civ. Engrs*, 1965, **31**, Aug., 337–360.
6. Anderson J.K. *et al.* Forth Road Bridge. *Proc. Instn Civ. Engrs*, 1965, **32**, Nov., 321–512.
7. Middleboe S. and Ferguson H. Closure confusion mars Severn decision. *New Civ. Engr*, 1984, 16 Feb., 4, 5 and 14.
8. The Royal Edward Dock at Avonmouth. *Engineer, Lond.*, 1908, **106**; 3 July, 7–19; 10 July, 33–35; 17 July, 66–67, 70–72; 24 July, 95–96.
9. Mackenzie J.B. The Avonmouth Dock. *Min. Proc. Instn Civ. Engrs*, 1878–79, Part 1, **55**, 3–21.
10. Irwin-Childs F. *et al.* The Royal Portbury Dock, Bristol. *Proc. Instn Civ. Engrs*, Part 1, 1978, **64**, Feb., 63–82.
11. Mallory K. *Clevedon Pier.* Redcliffe Press, Bristol, 1981.
12. Grover J.W. Description of a wrought-iron pier at Clevedon, Somerset. *Min. Proc. Instn Civ. Engrs*, 1870–71, Part 2, **32**, 130–136.
13. Opening of the pier at Clevedon, Somersetshire. *Ill. Lond. News*, 1869, 10 Apr., 369–370.
14. The new pier at Weston-super-Mare. *Ill. Lond. News*, 1867, 15 June, 600–601.
15. Porter Goff R.F.D. Brunel and the design of the Clifton Suspension Bridge. *Proc. Instn Civ. Engrs*, Part 1, 1974, **56**, Jan., 303–321.
16. Pugsley Sir A. (ed.). *The works of Isambard Kingdom Brunel.* Institution of Civil Engineers and University of Bristol, London and Bristol, 1976, 51–68.
17. Barlow W.H. Description of the Bristol Suspension Bridge. *Min. Proc. Instn Civ. Engrs*, 1866–67, **26**, 243–257.
18. Hadfield C. and Skempton A.W. *William Jessop, Engineer.* David and Charles, Newton Abbot, 1979, 222–242.
19. Buchanan R.A. I.K. Brunel and the Port of Bristol. *Trans. Newcomen Soc.*, 1969–70, **42**, 41–56.
20. Corlett E.G.B. The Steamship Great Britain. *Trans. R. Soc. Nav. Archit.*, 1971, **113**, 411–437.
21. Corlett E.G.B. *The iron ship.* Moonraker Press, Bradford-on-Avon, 1975.

22. *Job record sheet, No. 425.* J.T. Building Group Ltd, Bristol, 1976.
23. Green J. Account of recent improvements in the drainage and sewerage of Bristol. *Min. Proc. Instn Civ. Engrs*, 1848, **7**, 76–84.
24. Binnie G.M. *Early Victorian water engineers.* Thomas Telford, London, 1981, 82–91.
25. Coysh A.W. *et al. The Mendips.* Robt Hale, London, 1977, 196–201.
26. Simpson J. Discussion: Bridge-aqueduct at Roquefavour, by G. Rennie. *Min. Proc. Instn Civ. Engrs*, 1854–55, **14**, 218–223.
27. Pugsley Sir A. (ed.). *The works of Isambard Kingdom Brunel.* Institution of Civil Engineers and University of Bristol, London and Bristol, 1976, 121–122.
28. Brunel I.K. *Calculation book.* 1837. (Library, University of Bristol.)
29. Brunel I.K. *Private letter book.* 10 Oct. 1836. (Library, University of Bristol.)
30. Brunel I.K. Discussion: The construction of the collar roof with arched trusses of bent timber at Horsley Park, by the Earl of Lovelace. *Min. Proc. Instn Civ. Engrs*, 1849, **8**, 285–286.
31. Totterdell J. A peculiar form of construction. *J. Bristol & Somerset Soc. Archit.*, 1960, **5**, Mar., 111–112.
32. Barbey M.F. *Civil engineering heritage: northern England.* Thomas Telford, London, 1981, 60–61.
33. Barrie D.S.M. The Bristol and South Wales Union Railway. *Rly Mag.*, 1936, **79**, Dec., 423–427.
34. The Bristol and South Wales Union Railway. *Ill. Lond. News*, 1863, 5 Sept., 245.
35. Cole S.D. *The sea walls of the Severn.* Privately printed, Bristol, 1912.
36. Walker T.A. *The Severn Tunnel; its construction and difficulties,* 1872–1887. Bentley, London, 1891; Kingsmead Reprints, Bath, 1969.
37. Quoted by Webb S. and Webb B. *English local government: the story of the King's highway.* Longmans, 1913; reissued Frank Cass, London, 1963, 184.
38. McAdam J.L. *Narrative of affairs of the Bristol District of Roads from 1816 to 1824.* Bristol, 1825. (Bristol Record Office; Bristol Public Library.)
39. McAdam J.L. *Remarks on the present system of roadmaking.* Longman, London, 1827.
40. Cunliffe B. *Roman Bath discovered.* Routledge and Kegan Paul, London, 1984, 109–148.
41. Mann R. Note on a Roman culvert. *Proc. Br. Archaeolog. Ass.*, 1878, **34**, 15 May, 246–248.
42. Irvine J.T. (Discovery of a Roman reservoir at Bath.) *Proc. Br. Archaeolog. Ass.*, 1882, **38**, 4 Jan., 91–93.
43. Greenhalgh F. *Bath flood prevention scheme.* Wessex Water Authority, Bath, 1974.
44. Desarguliers J.T. *A course of experimental philosophy.* Printed for J.

Senex *et al.*, London, 1734, **1**, 274 – 279 with 3 plates.
45. Danes M. *The Silbury treasurer*. Thames and Hudson, 1976, chapters 1 – 3.

Chapter 5: Gloucestershire and Hereford and Worcester

1. Heyman J. and Threlfall B.D. Two masonry bridges: II – Telford's bridge at Over. *Proc. Instn Civ. Engrs*, Part 1, 1972, **52**, Nov., 319 – 330.
2. Telford T. (Rickman J. (ed.)). *Life of Thomas Telford . . . with a folio atlas of copper plates*. London, 1838, plate 66.
3. The new bridge at Over. *Glos. J.*, 1828, 24 May, 3.
4. Binnie G.M. *Early Victorian water engineers*. Thomas Telford, London, 1981, 173 – 174.
5. Brewster R.S. *The Port of Gloucester its history and development from the earliest settlement to 1976*. (Typescript 1978 in Gloucester Public Library.)
6. Conway-Jones A.H. *Gloucester Docks, an illustrated history*. Alan Sutton and Gloucester County Library, 1984.
7. Conway-Jones A.H. The warehouses at Gloucester Docks. *J. Glos. Soc. Ind. Archaeol.*, 1977 – 78, 13 – 19.
8. Owen G.W. Discussion: The different modes of erecting iron bridges, by T. Seyrig. *Min. Proc. Instn Civ. Engrs*, 1880 – 81, Part 1, **63**, 189.
9. Paar H.W. *The Severn and Wye Railway*. David and Charles, Newton Abbot, 1973, 102 – 111.
10. Huxley R. *The rise and fall of the Severn Bridge Railway, 1872 – 1970*. Alan Sutton and Gloucestershire County Library, 1984.
11. Handford M. *The Stroudwater Canal*. Alan Sutton, Gloucester, 1979.
12. Tann J. *Gloucestershire woollen mills*. David and Charles, Newton Abbot, 1967, 149 – 151.
13. Household H. *The Thames and Severn Canal*. Alan Sutton, Gloucester, 1983, 2nd edn.
14. Maggs C.G. *The Bristol and Gloucester and Avon and Gloucestershire Railways*. Oakwood Press, Lingfield, 1969.
15. Berridge P.S.A. *The girder bridge*. Maxwell, London, 1969, 82 – 136.
16. Mackenzie W. Account of the bridge over the Severn, near the town of Tewkesbury. *Trans. Instn Civ. Engrs*, 1838, **2**, 1 – 14.
17. Telford T. *Account of Tewkesbury Bridge*. Unpublished manuscript, Institution of Civil Engineers (Original Communication No. 126, 11 Mar. 1828).
18. Hammond B.C. The strengthening of a cast-iron bridge by welded steel bars encased in concrete. *Sel. Engng Pap. Instn Civ. Engrs*, 1934, No. 162, 3 – 15.

Chapter 6: West Midlands and South Staffordshire

1. Barbey M.F. *Civil engineering heritage: northern England*. Thomas Telford, London, 1981, 40 – 41.

2. Barbey M.F. *Civil engineering heritage: northern England.* Thomas Telford, London, 1981, 137 – 143.
3. Middleboe S. Repairers wary in Netherton heaving invert. *New Civ. Engr*, 1983, 30 June, 26 – 27.
4. Barbey M.F. *Civil engineering heritage: northern England.* Thomas Telford, London, 1981, 162 – 164.
5. Webster N.W. *Britain's first trunk line, the Grand Junction Railway.* Adams and Dart, Bath, 1972.
6. Roscoe T. *Book of the Grand Junction Railway.* Orr and Co., London, 1839.
7. Barbey M.F. *Civil engineering heritage: northern England.* Thomas Telford, London, 1981, 150.
8. Telford T. (Rickman J. (ed.)). *Life of Thomas Telford . . . with a folio atlas of copper plates.* London, 1838, plate 82.

Chapter 7: Shropshire

1. Mansergh E.L. and Mansergh W.L. The works for the supply of water to the City of Birmingham from Mid-Wales. *Min. Proc. Instn Civ. Engrs*, 1911 – 12, Part 4, **190**, 34 – 38.
2. Lapworth H. The construction of the Elan Aqueduct: Rhayader to Dolau. *Min. Proc. Instn Civ. Engrs*, 1899 – 1900, Part 2, **140**, 235 – 248.
3. Mansergh E.L. and Mansergh W.L. The works for the supply of water to the City of Birmingham from mid-Wales. *Min. Proc. Instn Civ. Engrs*, 1911 – 12, Part 4, **190**, 43 – 44.
4. Telford T. Letter to unknown correspondent (undated). (Telford Collection, Ironbridge Gorge Museum.)
5. Cossons N. and Trinder B. *The Iron Bridge.* Moonraker Press, Bradford-on-Avon, 1979, 79 – 86.
6. Tredgold T. *Practical essays on the strength of cast iron.* Neale, London, 1842, 2nd edn, 10.
7. Cossons N. and Trinder B. *The Iron Bridge.* Moonraker Press, Bradford-on-Avon, 1979, 11 – 36 and 48 – 52.
8. Cossons N. and Trinder B. *The Iron Bridge.* Ironbridge Gorge Museum Trust and Moonraker Press, Bradford-on-Avon, 1979, 119 – 125.
9. Skempton A.W. and Johnson H.R. The first iron frames. *Archit. Rev.*, 1962, **131**, 175 – 186.

Chapter 8: Cheshire and North Staffordshire

1. Telford T. Letter to Andrew Little of Langholm from Shrewsbury, 29 Sept. 1793. (Copy in Telford Collection, Ironbridge Gorge Museum.)
2. Bache A. Correspondence: The aesthetic treatment of bridge structures, by J. Husband. *Min. Proc. Instn Civ. Engrs*, 1900 – 01, **145**, 217.
3. Hartley J. An account of the new or Grosvenor Bridge over the River Dee at Chester. *Trans. Instn Civ. Engrs*, 1836, **1**, 207 – 214.

4. *British bridges.* Public Works, Roads and Transport Congress. London, 1933, 33–34.
5. Séjourné P. *Grandes voûtes.* Tardy-Pigelet, Bourges, 1913–16; **2**, 67–82; **3**, 29–31, 52–58 and 141–149.
6. Institution of Civil Engineers. Memoir on G. Rennie. *Proc. Instn Civ. Engrs*, 1868–69, **28**, 611–612.
7. Barbey M.F. *Civil engineering heritage: northern England.* Thomas Telford, London, 1981, 104.
8. Husband H.C. The Jodrell Bank radio telescope. *Proc. Instn Civ. Engrs*, 1958, **9**, Jan., 65–86.
9. Barbey M.F. *Civil engineering heritage: northern England.* Telford, London, 1981, 164–165.
10. Saner J.A. Swing-bridges over the River Weaver at Northwich. *Min. Proc. Instn Civ. Engrs*, 1899–1900, Part 2, **140**, 72–108.
11. Duer S. Hydraulic canal lift at Anderton, on the River Weaver. *Min. Proc. Instn Civ. Engrs*, 1875–76, Part 3, **45**, 107–129.
12. Wells L.B. Correspondence: Some canal, river, and other works, in France, Belgium and Germany, by L.F. Vernon Harcourt. *Min. Proc. Instn Civ. Engrs*, 1888–89, Part 2, **96**, 223–226.
13. Reconstruction of the Anderton boat lift. *Engineer, Lond.*, 1908, **106**, 24 July, 82–84 and 92.
14. Barbey M.F. *Civil engineering heritage: northern England.* Thomas Telford, London, 1981, 137–143.
15. Reed B. *Crewe locomotive works and its men.* David and Charles, Newton Abbot, 1982.
16. Barbey M.F. *Civil engineering heritage: northern England.* Thomas Telford, London, 1981, 161.
17. Barbey M.F. *Civil engineering heritage: northern England.* Thomas Telford, London, 1981, 164.
18. Barbey M.F. *Civil engineering heritage: northern England.* Thomas Telford, London, 1981, 34.
19. Anderson J.K. Runcorn-Widnes Bridge. *Proc. Instn Civ. Engrs*, 1964, **29**, Nov., 535–570.

Additional sites

North Wales

Aberffraw Bridge, HEW 163,
SH 355 689

Bangor Pier, HEW 427,
SH 584 732

Dinorwig Tramroad & Padarn
Railway, HEW 1286, SH 526
678 to SH 59 60

Llangollen Chain Bridge,
HEW 727, SJ 199 432

Lledr Viaduct (Gethin's Bridge),
HEW 1303, SH 780 539

Maentwrog Dam, HEW 399,
SH 673 377

Victoria Pier, Colwyn Bay,
HEW 1285, SH 853 792

Mid-Wales

Afon Rheidol Dams, HEW 1286,
SN 68 78, SN 74 82, SN 75 88

Berriew Aqueduct, HEW 1465,
SJ 189 006

Brithdir Aqueduct, HEW 522,
SJ 198 022

Buttington Bridge, HEW 852,
SJ 246 089

Claerwen Dam, HEW 186,
SN 86 63

Penarth Weir, River Severn,
HEW 1279, SO 140 729

Vyrnwy Canal Aqueduct,
HEW 1112, SJ 254 196

South Wales

Bridgend Old Town Bridge,
HEW 1031, SS 904 798

Bryn Tramroad Bridge, near
Port Talbot, HEW 1045, SS
787 923

Cenarth Bridge, HEW 166,
SN 269 415

Leckwith Ancient Bridge,
HEW 161, ST 851 915

Llwchwr Road Viaduct,
HEW 1231, SS 562 981

Maesteg Bridge, HEW 525,
SS 851 915

Merthyr (Ynysgau) Bridge,
HEW 805, SO 047 062

Pont Rhyd y Fen Bridge,
HEW 1004, SS 796 942

Pont Spwdwr Road Bridge,
HEW 1218, SN 434 059

Avon and North Wiltshire

Almondsbury Motorway
Interchange, HEW 49,
ST 617 837

Bradford-on-Avon Wharf, Kennet
and Avon Canal, HEW 383,
ST 825 601

Brislington Bridge, Bristol,
HEW 1143, ST 617 727

Cheltenham Road Railway
Bridge, Bristol, HEW 99,
ST 589 746

Cleveland Bridge, Bath,
HEW 650, ST 753 657

Limpley Stoke Viaduct,
HEW 377, ST 781 620

Maud Heath's Causeway near
Chippenham, HEW 680,

ST 972 737 to ST 919 739
North Parade Bridge, Bath,
HEW 339, ST 754 647
Priston Mill, HEW 1209,
ST 695 615
Severn Cable Tunnel,
HEW 1144, ST 555 902

Gloucestershire and Hereford and Worcester
Berkeley Nuclear Power Station,
HEW 1181, ST 660 995
Bibury Bridge, HEW 1140,
SP 115 068
Bringewood Forge Bridge,
HEW 1278, SO 454 750
Bullo Pill, HEW 1332,
SO 690 099
Danzey Green Post Mill (at
Avoncroft Museum),
HEW 864, SO 454 750
Donnington Brewery, near
Stow-on-the-Wold, HEW 1271,
SP 174 272
Droitwich Canal, HEW 183,
SO 842 599 to SO 904 635 to
SO 422 629
Hewlett's Reservoirs,
Cheltenham, HEW 1308,
SO 97 22
Masonry Bridge, Cinderford,
HEW 1311, SO 642 125
Redbrook Incline, Monmouth
Tramroad, HEW 627,
SO 537 102
Stourport Bridge, HEW 1051,
SO 808 711
Tintern Tramway Bridge,
HEW 930, SO 530 003
Witcombe Reservoir, near
Gloucester, HEW 1309,
SO 90 15

West Midlands and South Staffordshire
Anker Viaduct, Tamworth,
HEW 886, SK 213 036

Birchills Canal Aqueduct,
HEW 830, SK 009 004
Brownhills Aqueduct, HEW 283,
SK 053 064
Coat of Arms Bridge, Coventry,
HEW 1030, SP 325 767

Shropshire
Atcham Old Bridge, near
Shrewsbury, HEW 732,
SJ 541 093
Attingham Park Lock,
Shrewsbury, HEW 638,
SJ 553 092
Boreton (Cliff) Cast Iron Bridge,
Cound Brook, HEW 978,
SJ 517 068
Borle Brook Footbridge,
HEW 977, SO 753 817
Bridgnorth Station, HEW 640,
SO 715 926
Greyfriars Footbridge,
Shrewsbury, HEW 1104,
SJ 495 121
Hadley Park Lock, Shropshire
Union Canal, HEW 938,
SJ 672 132
Hampton Loade, Old Forge
Bridge, HEW 971, SO 747 863
Madeley Church, HEW 1333,
SJ 696 041
Porthill Footbridge, Shrewsbury,
HEW 1102, SJ 485 126

Cheshire and North Staffordshire
Bollinhurst Dam, HEW 1151,
SJ 973 836
Bottoms Dam, HEW 1153,
SJ 945 716
Congleton Aqueduct, Macclesfield
Canal, HEW 494, SJ 866 622
Congleton Viaduct, HEW 1336,
SJ 877 627
Dane Aqueduct, Macclesfield
Canal, HEW 1343, SJ 906 652
Horse Coppice Dam, HEW 1152,

SJ 968 837

Moving Old Academy Building,
   Warrington, HEW 1085,
   SJ 607 879

North Rode Viaduct, HEW 519,
   SJ 896 657

Prestressed concrete arch railway
   underbridges, Runcorn area,

HEW 259, SJ 533 797 and
   SJ 554 797

Queensferry Bridge, HEW 1317,
   SJ 322 687

Rudyard Dam, Leek, HEW 1472,
   SJ 951 583

Stoke-on-Trent Station roof,
   HEW 764, SJ 879 456

Index of engineers, architects and contractors

Subject index